D0848634

Springer Series in
MATERIALS SCIENCE 59

Springer
Berlin
Heidelberg
New York
Hong Kong
London
Milan
Paris
Tokyo

Physics and Astronomy

ONLINE LIBRARY

http://www.springer.de/phys/

Springer Series in
MATERIALS SCIENCE

Editors: R. Hull   R. M. Osgood, Jr.   J. Parisi

The Springer Series in Materials Science covers the complete spectrum of materials physics, including fundamental principles, physical properties, materials theory and design. Recognizing the increasing importance of materials science in future device technologies, the book titles in this series reflect the state-of-the-art in understanding and controlling the structure and properties of all important classes of materials.

Series homepage – http://www.springer.de/phys/books/ssms/

Volumes 1–50 are listed at the end of the book.

S. O. Gladkov

Dielectric Properties
of Porous Media

With 71 Figures

 Springer

Professor S. O. Gladkov
Moscow Pedagogical University
Chair of Mathematical Analysis
Ul. Radio 10A
107005 Moscow, Russia

Series Editors:
Professor Robert Hull
University of Virginia, Dept. of Materials Science and Engineering, Thornten Hall
Charlottesville, VA 22903-2442, USA

Professor R. M. Osgood, Jr.
Microelectronics Science Laboratory, Department of Electrical Engineering
Columbia University, Seeley W. Mudd Building, New York, NY 10027, USA

Professor Jürgen Parisi
Universität Oldenburg, Fachbereich Physik, Abt. Energie- und Halbleiterforschung
Carl-von-Ossietzky-Strasse 9-11, 26129 Oldenburg, Germany

ISSN 0933-33X
ISBN 3-540-00186-7 Springer-Verlag Berlin Heidelberg New York

Cataloging-in-Publication Data applied for
Bibliographic information published by Die Deutsche Bibliothek.
Die Deutsche Bibliothek lists this publication in the Deutsche Nationalbibliografie;
detailed bibliographic data is available in the Internet at<http://dnb.ddb.de>.

Springer-Verlag Berlin Heidelberg New York
a member of BertelsmannSpringer Science + Business Media GmbH

http://www.springer.de

© Springer-Verlag Berlin Heidelberg 2003
Printed in Germany

Typesetting: PTP, Berlin
Cover concept: eStudio Calamar Steinen
Cover-Design: design + production, Heidelberg
Printed on acid-free paper   57/3020/hu 5 4 3 2 1 0

To my near relations and my people

Preface

This monograph is the first systematic presentation of the theoretical fundamentals of the physics of porous structures and it is intended for basic familiarization with the subject. That is why we tried to avoid, where possible,
excessive mathematics and to pay as much attention to the visual presentation of the general physical picture of various phenomena as possible. Even
in those cases where there is an accurate mathematical theory we avoid rigorous mathematical proofs, assuming that the reader is acquainted with the
fundamentals of quantum mechanics and statistical physics and in particular
cases will be able to reproduce the calculations of interest.

For a more detailed and sound study of particular problems the reader
may refer to the original papers and monographs listed at the end of the
book.

The book includes five chapters. The first chapter deals with the model
of porous medium based on the notion of the fibril structure of substance
which allows us theoretically analyze some seemingly "desperate" problems.
Moreover, included in this chapter is some information from the thermodynamics of equilibrium systems. The second chapter deals with a number of
issues connected with the investigation of purely equilibrium characteristics
of such structures, including heat capacity and mechanical strength. Chapter 4 concerns the physical bases of fast processes in porous media. Chapter 5
shows the development of a consistent theory of heat conduction in porous
substances.

Finally, the last chapter, Chap. 6, includes analytical calculations of various kinetic parameters of porous dielectrics exposed to external alternating
fields as well as an explanation of some experiments.

The author would like to express his special gratitude to A.M. Tokarev
for conducting a number of time-consuming experiments on measuring some
important characteristics of porous substances, such as heat capacity, sound
velocity, ignition delay time, and thermal conductivity. The author is especially thankful to Dr. A.L. Bugrimov for the mathematical program which
provided the computer calculation of the coefficient of thermal conductivity.
Special thanks goes to Dr. I.V. Gladyshev for a professional description of
some theoretical results (heat conductivity coefficient of porous crystalline

dielectrics) using numerical methods. I am grateful for their patience and devotion.

I would like to thank my colleges Drs. V.G. Nikol'skii, N.Ya. Kuznetcova and Ms. I.A. Krasotkina for discussing some scientific results.

Moscow, February 2003   *S.O. Gladkov*

Contents

Nomenclature

A	wave function of photon (V)
B	vector of magnetic induction (Tesla, or Vs/m^2)
c_P	isobaric heat capacity per unit volume (m^{-3})
c_V	isochoric heat capacity per unit volume (m^{-3})
C_P	isobaric heat capacity
C_V	isochoric heat capacity
c	light speed in vacuum (3×10^8 m/s)
c_s	average sound speed in substance (m/s)
c_l	longitudinal speed of sound (m/s)
c_t	transverse speed of sound (m/s)
d	distance between pores (m)
$\boldsymbol{d}$	dipole moment of atom (Vm^2)
$\boldsymbol{D}$	vector of electrical induction (V/m)
e	charge of electron 1.602×10^{-19} (C)
$\boldsymbol{E}^{(i)}$	internal electrical field (V/m)
$\boldsymbol{E}^{(e)}$	external electrical field (V/m)
$\boldsymbol{E}_0$	constant and homogenous electric field (V/m)
E	internal energy (J)
f	distribution function
f_0	quasi-equilibrium distribution function of molecules
$\boldsymbol{F}$	force on the sample (J/m)
F	Helmholtz free energy (J)
$\boldsymbol{g}$	gravitational acceleration vector (m/s^2)
g	gravitational acceleration (m/s^2)
$\boldsymbol{h}$	radio-frequency magnetic field (Vs/m^2)
$\boldsymbol{H}^{(i)}$	internal magnetic field (Vs/m^2)
$\boldsymbol{H}^{(e)}$	external magnetic field (Vs/m^2)
$\boldsymbol{H}_0$	constant and homogenous magnetic field (Vs/m^2)
h	extinction coefficient (m^{-1})
h_{12}	depth of penetration of electromagnetic radiation into substance in phonon–photon interaction (m)
h_{21}	depth of penetration of electromagnetic radiation into substance in photon–phonon interaction (m)
H	Hamiltonian (J)

$i, k, \ldots,$ summation indices
j current density (A/m^2)
J intensity of radiation (W/m^2)
k_B Boltzmann's constant: 1.381×10^{-23} (J/K)
k_k Kozeny's constant
K permeability coefficient (m^2)
l length of free path of quasi-particle (m)
m porosity
m_e mass of electron: $0.9109534 \times 10^{-30}$ (kg)
M mass of elementary cell (kg)
M_m mass of molecule (kg)
P pressure (Pa)
q flux of heat (W/m^2)
Q quantity of heat (J)
r radial coordinate (m)
R radius (m)
s entropy per unit volume (m^{-3})($k_B \equiv 1$)
S entropy
t time (s)
T temperature (K)
V speed of molecular flux (m/s)
x, y, z coordinate axis (m)
W enthalpy (J)

Greek

α heat irradiation coefficient (J m^{-2}s^{-1})
$\alpha, \beta \ldots,$ summation indices
γ attenuation coefficient (s^{-1})
δ thickness (m)
Δ Laplace's operator (m^{-2})
ε dielectric permeability
η viscosity (kg m^{-1}s^{-1})
θ polarity angle (rad)
κ conductivity (W m^{-1}s^{-1})
λ wavelength of particle (m)
μ magnetic permeability
χ temperature conductivity coefficient (m^{2}s^{-1})
ρ density (kg m^{-3})
σ Stefan-Boltzmann constant: 5.6696×10^{-8} (W m^{-2}K^{-4})
σ_e electrical conductivity (who m^{-1})
φ electrical potential (V)
Φ Gibbs potential (J)
ω frequency (rad s^{-1})
Ω solid angle (sr)

Superscripts

—	equilibrium value
$*$	dimensionless value
$'$	derivative
$''$	second derivative
int	interaction
$\wedge$	operator
0	interaction is absent

Subscripts

D	Darcy
K	Kozeny
T	temperature

Others

$\lVert \dots \rVert$	matrix
$(\dots)$	private derivative

1 Introduction

In the classic monograph by Gees [1] the chemical nature of cellulose, one of the most practically important porous substances, is thoroughly examined. Also dicussed in the monograph are the main microscopic dimensions of atomic bounds and complexes, and the molecular structure is described. The analysis of the photographs obtained by electron microscopy allows us to determine the internal macroscopic structure of such dielectrics and to determine the entire quantity of fibers in a sample, their average length and their internal and external diameters. More powerful magnification allows us to estimate the fibril dimensions.

Issuing from these and some other experimentally measured parameters we will be able to estimate numerically the effects with the view to finding out the qualitative behavior of the most important parameters of such non-ordered structures. Besides, accurate knowledge of the microscopic distances between hydroxyl groups and "the length" of the intermolecular bonds enables us to make, after the corresponding theoretical calculations, the necessary estimation and comparison with the experiments (though not numerous in this field) carried out to measure the "delta tangent" or the electric breakdown field and its functional dependencies.

As we shall see later, there is quite satisfactory agreement with experiment, which, in our opinion, does not give any grounds for pessimism. Some idealized concept of the porous structure of a substance as a quite ordered (on average) stucture turns out to be rather useful in purely theoretical aspects.

The first investigations of the properties of porous media may have been carried out at the beginning of the twentieth century, when various electromechanical devices (such as transformers) began to appear, insulating pads being their most important components. Experimental observations carried out in the 1920s and 1930s and aimed mainly at the study of purely technical characteristics, e.g. capacitor paper, became a powerful stimulus for the development of this new class of non-ordered structure. During various experiments paper sheets were impregnated with all kinds of oils and fluids in order to increase the electric strength of the items. Such experiments, which may be related to the 1950s or 1960s were conducted using organic materials of the glass or ceramic type. The most important characteristic of such structures is their porosity. It turns out that in studies of substances with homogeneous composition (here we do not mean such effects as phase transitions of the first and the second kind, the Kondo effect and the like, which are not connected with the structure porosity) no abnormal phenomena were observed. As soon as we add free volumes into the main matrix (modern technology allows us to do so), certain physical indices change qualitatively to a rather great extent.

Here the entropy of such an inhomogeneous substance becomes lower than a homogeneous one. Indeed, if we hypothetically cut that out of of the structure one free volume, the number of atoms composing its basis will decrease, and due to it the entropy will decrease as well (it is known that the entropy is directly – but not proportionally! – connected with the number of elementary units of a structure).

However, in spite of the fact that the system on the whole decreased its entropy and lost some energy (the internal energy increased!) there appeared another possibility to compensate it due to gaining a number of abnormal properties not characteristic of a homogeneous substance with abnormal properties. Such compensation is very similar to the principle of Le Chatelier.

The subject under investigation is in itself rather complicated because of the chaotic arrangement of voids in the volume. It is clear that the calculation of the physical parameters of such a non-ordered structure is very complicated theoretically; that is why Chap. 2 is devoted to a hypothetical model. The approach used as the basic one may be justified only in the case where the predicted theoretical results will not be in contradiction with the empirical and numerical estimations known from the experiment.

Further, we shall see that the fibril model, which is a kind of "reference point", correctly corresponds to the experimental results and, in our opinion, is a convincing proof of the possibility of its application to the interpretation of certain experimental data. As to the application of the model in question to a more complex substance, quite satisfactory agreement may be achieved by the introduction of corresponding correction factors, implying that the system is not ideal.

It should be also stressed that the variety of porous structures is not exhausted by paper and cellulose, other such structures include porous crystal dielectrics, porous magnetic materials, spin glasses, metals and alloys, and finally great number of other natural (solidified magma, timber, all the cosmic bodies including the Earth, ...) and artificial (rubber sponge, aerated water, concrete, brick, ...) substances having (just due to the presence of pores) unique properties.

From the above mentioned substances magnetic and metallic porous media are the most curious, and is not due to the fact that the main matrix is magnetized or possesses conductive properties, (though this is not unimportant!) but rather due to the combination of different phases (in physical characteristics) and, most importantly, due to their bilateral interaction!

As a concrete example we shall choose a metal heterogeneous structure and see how pores influence some properties.

Free electrons from the conduction band of the main matrix must accumulate on one side of the pores forming a negative charge, while positive (hole) charges will be localized on the opposite side. This means that the gas phase is continuously exposed to the action of an internal constant field, and therefore the sample itself on the whole is in a stressed striction state all the time.

Suppose now, that such a sample is put into an external constant magnetic field. As we know from any general course of physics, a free electron must "twist" on a spiral path around the isolated direction of the magnetic field with some constant pitch. What will happen in our case? In our case electrons (if the magnitude of the magnetic field is rather high) will start to describe concentric circles of variable radii along the pore surfaces on a spiral path "flowing off" inside the volume and "dissolving" in it .It is for this reason that such important physical characteristics as magnetic susceptibility and permeability must be defined not so much by "volume" electrons as by "surface" ones, despite the fact that their concentration, proportional to the quantity of pores, is much lower. Just at this point a very delicate problem arises connected with the way to calculate the magnetic susceptibility coefficient, taking into account "surface" electrons.

One more example: Let us now take a magnet dielectric with a single spherical pore of radius R and inquire, how pore size and properties of the gas filling it influence on the thermal conductivity of the whole magnet dielectric in general. In order to answer this question, first we should recall magnetic properties and introduce a quasi-particle, magnon. Unlike the classical electron, a magnon has no mass and may be viewed only as a wave; that is why a concept of dispersion was introduced for it , defining the dependency of its frequency on the wave vector k. As in the magnetic volume there is inhomogeneity in the form of a "void" sphere, the problem arises: how does the law of magnon dispersion change as a result of violation by such homogeneity of translational invariance of the whole crystal? Far from being trivial, such a problem forma the basis of a first approach to the solution of the task of calculating the thermal conductivity coefficient of a porous magnetic substance.

The other approach is to consider separately the thermal conductance properties of a pure magnet and a pore filled with a gas or a liquid, but taking into account their "exchange" (not in magnetic terms, that is why we have written the word in adverted commas) interaction in some transition region of width δ, separating both phases with a relatively sharp boundary. The latter approach has been developed and described in detail in the monograph [124].

In the manuscript presented to the reader, "Dielectric properties of porous media", we shall not delve deeper in detail into the analysis of the stated problems because they are beyond the subjects touched upon in the present monograph, but we shall only add that both problems are very complicated ones and for their solution it is necessary to possess knowledge not only of quantum mechanics and statistical physics, but it is also necessary to master some methods of theoretical physics!

Summing up our short introduction, we shall once more draw the attention of beginning researchers to a great number of unsolved but practically very important problems from the physics of heterogeneous structures that have been stated; now they are waiting for their turn!

2 Model of the Internal Structure of a Porous Dielectric

2.1 Ideal Model of a Porous Substance with Ideal Compact Packing

The principal topic of the present chapter may cause a feeling of distrust, and the reader will say at once that there is no ideal structure. Such a reader most probably belongs to the class of practical workers. But a theorist who realizes the difficulty of describing such structures will note that it is worthwhile considering such a model, but making it closer to reality. If it will help to explain at least some experimental results, then the suggested model has the right to exist. Note that model problems of such a type were solved and described in many publications both, Russian and foreign, and there is a great number of original papers and classical monographs [1–9].

To begin with, we shall describe in elementary language such a physically important concept as the density of porous structures. The structure discussed below is a chaotic weave of thin filaments and the space between them is filled with a gaseous phase. The filament itself may be both hollow and continuous. We deliberately do not discuss the atomic (or molecular) structure of the filaments, which we shall call fibers, because their microscopic structure will not be so important (though in some cases where microscopy will be necessary, we shall specially discuss the corresponding problem statements).

Thus, suppose that in the ideal case a fiber can be considered as a hollow cylindrical tube of length L and external radii R_1 and R_2, respectively. If we now take a cube with the edge length also equal to L and uniformly fill it with hollow tubes in parallel rows, we shall get the picture shown in Fig. 2.1a. In this case the density of this ideal structure is determined very easily. Indeed,

$$\rho = \frac{M}{V} = \frac{(M_f + M_p)}{V} = (1 - m)\rho_f + m\rho_p \,, \tag{2.1}$$

where ρ_f and ρ_p are the fiber and pore densities, respectively.

Here, the porosity is

$$m = \frac{V_p}{V} \,, \tag{2.2}$$

where V_p is the volume of the pores.

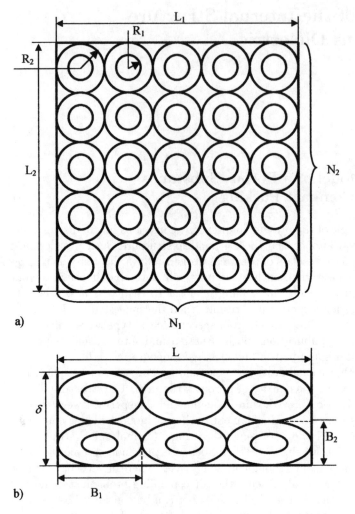

Fig. 2.1. Ideal packing of shallow fibers before mechanical action (**a**) and on exposure to pressure (**b**)

Because (see Fig. 2.1a) $V_p = V - V_f = L^3 - \pi(R_1^2 - R_2^2)LN$, where $N = N_x N_y$ is the entire number of hollow cylinders (fibers) in the cube section, the porosity is

$$m = \frac{L^3 - \pi R_1^2 - R_2^2)LN}{L^3} = 1 - \frac{\pi(R_1^2 - R_2^2)N}{L^2}. \tag{2.3}$$

Since $N = (L/2R_1)^2$, then $m = 1 - 0.25\,\pi(1 - R_2^2/R_1^2)$ and therefore the density of such a structure is

$$\rho = 0.25\,\pi\rho_f\left(1 - \frac{R_2^2}{R_1^2}\right) + \rho_p\left[1 - 0.25\,\pi\left(1 - \frac{R_2^2}{R_1^2}\right)\right]. \tag{2.4}$$

It is clear that a substance with the ideal arrangement of fibers would have such a density. Now let this substance be exposed to mechanical action – compression. After such exposure with the dimension L kept fixed (Fig. 2.1b) we obtain

$$\rho = \frac{\rho_f L B_1}{B_2 \delta} + \rho_p \left(1 - \frac{L B_1}{B_2 \delta}\right). \tag{2.5}$$

Because $\delta/B_1 = L/B_2 + \varepsilon$, where the parameter $\varepsilon \Rightarrow 0$ and formally characterizes the relative zone of the unfilled voids (sometimes called free volumes), we obtain approximately

$$\rho = (1 - \varepsilon_1)\rho_f + \varepsilon_1 \rho_p, \tag{2.6}$$

where $\varepsilon_1 = \varepsilon B_1/\delta$.

In order to approach still closer the real substance we shall consider the fiber to consist of separate smaller but continuous cylindrical particles in practice called fibrils. The free space between them is also filled with gas. Taking this into account, the density of the fiber is

$$\rho = (1 - \mu)\rho_{fl} + \mu\rho_p, \tag{2.7}$$

where ρ_{fl} is the fibril density and $\mu = v_{pf}/v_f$ is the porosity of the fiber. Therefore,

$$\rho = (1 - \mu)\rho_{fl} + \mu\rho_p \leq \rho_{fl}. \tag{2.8}$$

The equality sign in the above equation holds only in the ideal case when free volumes are absent absolutely. Since in reality this is never the case, we may believe that always

$$\rho \leq \rho_{fl}. \tag{2.9}$$

Indeed, substituting (2.7) into (2.6) and collecting similar terms we find

$$\rho = \rho_{fl}(1 - \mu)(1 - \varepsilon_1) + \rho_p[\varepsilon_1 + \mu(1 - \varepsilon_1)] \cong \rho_{fl}(1 - \mu)(1 - \varepsilon_1). \tag{2.10}$$

The above equation obtained enables us to conclude that the inequality (2.9) is always valid and that practically it is impossible to obtain the density of the structure with zero porosity. Note, by the way, that the fibril density for such substances as paper or cellulose is $\rho_{fl} \cong 1.55\,\mathrm{g/cm}^3$.

Let us now consider another ideal case when there are no fibers but only a fibril structure and the fibrils fill the cube volume uniformly (Fig. 2.2a). In this case the packing density is

$$\rho^* = \frac{M}{V} = \frac{\rho_{fl} v_{fl} N_{fl} + \rho(V - N_{fl} v_{fl})}{V}. \tag{2.11}$$

Assuming conditionally that the fibril length is equal to $l_{fl} = L/k$, where k is the number of fibrils in the fiber, we get

$$\rho^* = \frac{\rho_{fl} \pi r_{fl}^2 N_{fl} k^2 + \rho_p(L^2 - \pi r_{fl}^2 N_{fl} k^2)}{L^2}. \tag{2.12}$$

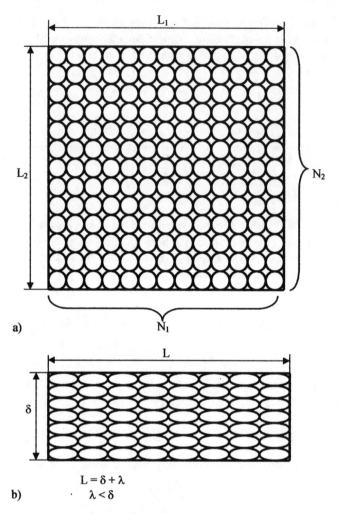

Fig. 2.2. Model representation of fibril ideal packing before compression (**a**) and after compression (**b**)

As the number of fibrils is

$$N_{fl} = N_1 N_2 = \left(\frac{L}{2r_{fl}} \right)^2 , \tag{2.13}$$

therefore

$$\rho^* = 0.25\pi\rho_{fl} + (1 - 0.25\pi)\rho_p . \tag{2.14}$$

The latter equation is valid if the substance is not exposed to mechanical action. After compression (e.g. rolling), according to Fig. 2.2b, the density will be as follows:

$$\rho^{**} = \frac{M}{V_0} = \frac{M}{L^2\delta} = \frac{\rho^* L}{\delta} \tag{2.15}$$

or with the account of (2.14), neglecting the addend containing ρ_p, we obtain the approximate equation:

$$\rho^{**} = 0.25\pi\rho_{fl}\frac{L}{\delta}. \tag{2.16}$$

Now, comparing the obtained equation with the dependence (2.10) we may conclude that the density of the purely fibril structure (after compression!) will be higher than that of the fiber structure only if

$$0.25\pi\frac{L}{\delta} > (1 - \varepsilon_1)(1 - \mu). \tag{2.17}$$

Let us consider one more kind of packing. It concerns so called two-dispersity filling, i.e. when there are both fibers and single fibrils. In this case, according to Fig. 2.3 we have

$$\langle\rho\rangle = \frac{M}{V} = \frac{\rho_f N_f v_f + \rho_{fl} N_{fl} v_{fl} + \rho_p (V - v_f N_f - v_{fl} N_{fl})}{V} \tag{2.18}$$

or, since

$$N_f = \left(\frac{L}{2R_1}\right)^2 \quad \text{and} \quad N_{fl} = \frac{L^3 - N_f v_f}{v_{fl}} = \frac{L^2(1 - 0.25\pi)}{\pi r_{fl}^2},$$

then after substitution into (2.18) we finally get

$$\langle\rho\rangle = 0.25\pi\rho_f\left(1 - \frac{R_2^2}{R_1^2}\right) + \rho_{fl}(1 - 0.25\pi) + 0.25\pi\rho_p\frac{R_2^2}{R_1^2}. \tag{2.19}$$

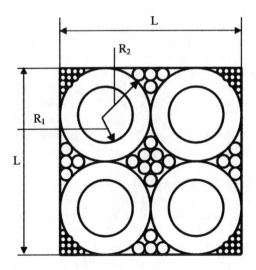

Fig. 2.3. Ideal two-dispersion phase consisting of fibers and fibrils

Compare now the densities ρ, ρ^* and $\langle\rho\rangle$ given by the formulae (2.4), (2.14) and (2.19), respectively.

In the case when the conditions in (2.17) are satisfied we see that the following relations take place:

$$\langle\rho\rangle > \rho^* > \rho. \tag{2.20}$$

Here it should be stressed that the relation $\langle\rho\rangle > \rho^*$ is valid practically in all the cases when the fiber wall is not too thin. After mechanical compression of such a hypothetical two-dispersity system and under the condition that the horizontal dimension L is constant, the density increases and in accordance with the formula (2.19), analogous to obtaining the relations (2.10) and (2.16), we find

$$\rho_0 = \langle\rho\rangle > \frac{L}{\delta}. \tag{2.21}$$

If now we compare the relations (2.10), (2.16) and (2.21), then unlike the inequality (2.20) we shall get the relation:

$$\langle\rho\rangle > \rho^{**} > \rho_0. \tag{2.22}$$

From the given inequality we come to a rather interesting conclusion that (at the not too high pressures created by the press) the fiber structure rather than the purely fibril one will have the highest density. Qualitatively, such a statement is quite clear. Indeed, due to the natural elasticity of fibers their walls are easily deformed and compressed into the core but the external surface of the fiber tends to fill those voids which have the common boundary with it.

In other words, the fiber, so to say, creeps in all directions. Since all the fibers behave in such a manner, the domain of free space will be maximally occupied. As far as the fibrils are concerned, they cannot have such elasticity due to their structure, and behave like solid bodies. It means that under the same mechanical loading their volume is deformed much more weakly than that of the fibers and the voids remain unfilled. Thus we may say that the density of the fibril structure practically did not change upon loading.

It is obvious that the above reasoning is correct only for the case when pressures are relatively low. If the pressure is high, as under conditions of a pulse local contact, the given local domain will have a density almost equal to the fibril density (as stated above, this value is $\rho_{fl} \cong 1.55$ g/cm^3). Therefore, the conclusion is definite: it is impossible to get higher structure density by macroscopic methods.

If we apply some other techniques that allow us to disintegrate the fibril structure at the molecular level, we shall get a solid body characterized by the usual exchange bonds between molecules and atom conglomerates rather than a porous substance.

A reasonable question may arise: can we increase the structure density (before the compression process!) just by decreasing the diameter of the particles filling up the volume? For it seems that with the decrease of their radius

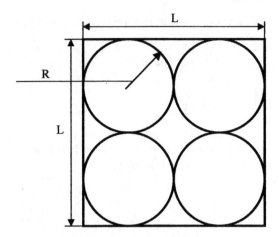

Fig. 2.4. Close packing of balls with radius R in the volume of the cube with L edge

they must occupy less volume. It does not appear to be so. Let us explain this statement.

Imagine a cube uniformly filled with ordinary balls (or continuous cylinders) with material of density ρ_b. Then from Fig. 2.4 it follows that the cube density is $\rho_c = M/V_c = N_b v_b \rho_b / V_c$, where $N_b = V_c / v_b$, and therefore $N_b v_b = V_c = \mathrm{const}$. Thus the decrease of the ball volume leads only to the decrease of their quantity, without affecting the density, because the product $N_b v_b$ remains unchanged and hence $\rho_c = \rho_b = \mathrm{const}$.

2.2 Influence of the Distribution of the Pore Volumes on the Structure Density

As we pointed out earlier, any porous medium has the highest density only if the pore volume becomes the lowest. In order to achieve it there is a number of technological cycles which provide uniform grinding of the substance with its subsequent dissolution and processing under pressure. As can be seen in practice [2–9], the highest density of the product is achieved only when the fibers and fibrils are laid uniformly and simultaneously (two-dispersity medium) but with some spread over the mean dimension. This may be shown in purely mathematical terms. Let us take the predetermined volume V_0 in which balls are closely packed with dimensions spread around some mean value v_0. Besides, we have the same value V_0 but packed with balls of two sorts (see Fig. 2.3 for the condition that $R_1 = 0$). One sort is the ensemble of balls with the volume spread near the mean value v_1 and the other near v_2. The number of balls of the first type is N_1 and of the second type N_2. In order to estimate which of these two types of packing has the higher density we shall calculate the densities of each of them and make a comparison. In the first case we have

$$\frac{1}{\rho_m} = \frac{V_0}{\Sigma M} = \frac{(V_p + V_B)}{M_0} = \frac{V_p}{M_0} + \frac{N \int_0^\infty v f(v) dv}{M_0 \int_0^\infty f(v) dv}, \tag{2.23}$$

where $\Sigma M = M_0$ is the entire mass of the balls, V_p is the volume of the pores, V_B is the volume of the balls, $f(v)$ is the density function of their distribution over volume and N is the number of balls in the volume V_0.

Assuming that $M_0/V_p = \Delta_p$, we find:

$$\frac{1}{\rho_m} = \frac{1}{\Delta\rho} + N \int \frac{v f(v) dv}{M_0 Z}, \tag{2.24}$$

where Z is the normalization integral:

$$Z = \int_0^\infty f(v) dv.$$

In order to concretize the results, let us assume that $f(v)$ is described by the Poisson distribution, that is:

$$f(v) = \left(\frac{v}{v_0}\right)^n \frac{\exp\left(-\frac{v}{v_0}\right)}{n! v_0^*}, \tag{2.25}$$

where v_0^* is some average volume and (2.24) gives

$$\frac{1}{\rho_m} = \frac{1}{\Delta\rho} + \frac{N v_0 (n-1)}{M_0}. \tag{2.26}$$

Now we have at our disposal balls of two types. Then

$$\frac{1}{\rho_m^*} = \frac{1}{\Delta\rho} + \frac{N_1 \int v f_1(v) dv + N_2 \int v f_2(v) dv}{N_1 m_1 \int f_1(v) dv + N_2 m_2 \int f_2(v) dv}, \tag{2.27}$$

where $m_{1,2}$ is the mass of one ball of the first and second sort, respectively; $\Delta\rho = M_0/V_p$ in the second case as well as in the first case. The entire mass of the balls M_0 may be considered equal in both cases.

Following this sequence of reasoning let us suppose that

$$f_{1,2}(v) = f(v) = \left(\frac{v}{v_{1,2}}\right)^n \frac{\exp\left(-\frac{v}{v_0}\right)}{n! v_0^*},$$

and therefore

$$\frac{1}{\rho_m^*} = \frac{1}{\Delta\rho} + \frac{(v_1^2 N_1 + v_2^2 N_2)(n-1)}{N_1 m_1 v_1 + N_2 m_2 v_2}. \tag{2.28}$$

In order to estimate in what case the density will be higher, we now shall obtain the difference between (2.26) and (2.28). Indeed, we have

$$\Delta = \frac{1}{\rho_m^*} - \frac{1}{\rho_m} = \left[\frac{(v_1^2 N_1 + v_2^2 N_2)}{(M_1 v_1 + M_2 v_2)} - \frac{N v_0}{M} \right] (n-1) , \tag{2.29}$$

where $M_1 = N_1 m_1$, $M_2 = N_2 m_2$.

Since $N > N_1 + N_2$ and $M = M_1 + M_2$, assuming, for the sake of accuracy, for example, that $v_0 = v_1$ we get

$$\Delta = \left[\frac{v_1 v_2}{(M_1 v_1 + M_2 v_2)} \right] + \left[\frac{N_1}{\xi} + N_2 \xi - N \left(\frac{\eta}{\xi} + 1 - \eta \right) \right] , \tag{2.30}$$

where $\xi = v_2 / v_1 > 1 (v_2 > v_1)$ and the parameter $\eta = M_1 / (M_1 + M_2) < 1$.

Consider now the equation in brackets. Suppose that $\Delta < 0$, then

$$N_1 + N_2 \xi^2 - N\xi(1 - \eta) - N\eta < 0 . \tag{2.31}$$

Solving the obtained inequality relative to the parameter ξ at obvious additional terms: $N_1 > N_2$, $M_2 > M_1$, we find that

$$1 < \xi < \frac{N(1-\eta)}{2N_2} + \left[\frac{N^2(1-\eta^2)}{4N_2^2} + \frac{(N\eta - N_1)}{N_2} \right]^{1/2} . \tag{2.32}$$

Here we stress that when $0.5 < \eta < 1$, the difference $N\eta - N_1$ is always positive.

Solving, at last, the inequality (2.32) we get the relation $N > N_1 + N_2$. Thus we have found out that the following relation is always valid:

$$\rho_m^* > \rho_m , \tag{2.33}$$

which had to be proved. Therefore, the structure density is the highest when we use the fiber and fibril type of packing.

Since our ultimate interest is focused on a highly disordered (geometrically) substance, the introduction of such an important concept as the elementary cell is caused not so much by the necessity to simplify theoretical calculations as by the wish to simplify the modeling of processes taking place in structures of this class. Note that the development of computer programs in this sphere will allow us to describe adequately any important technical characteristics of such highly disordered systems. The correct theoretical description, even based on "common sense", will not allow us to construct a model which would be able to describe uniformly some physical parameters of a porous dielectric without considering the cell structure.

The three possible types of the substance-packing analyzed in the previous section include in their structure fibril "grains" as a necessary element. The existence of such components leads to the question of wether it is possible to try to use fibrils as an elementary particle of the structure in order to construct some hypothetical elementary cell. In other words, wether it is possible to try to construct a cell only from fibrils packed in parallel and

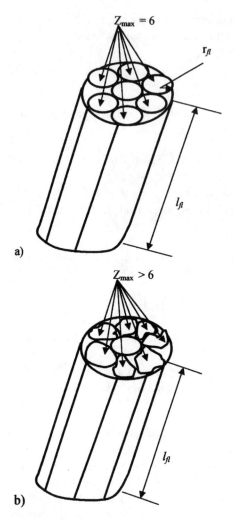

Fig. 2.5. Ideal "elementary cell" composed of cylindrical fibrils with length l_{fl} and radius r_{fl} (**a**) and approximation of the "elementary cell" to reality (**b**) ($z = 6$, $z_{\max} = 9$)

connected by the exchange interaction whose function is performed by van der Waals forces.

In Fig. 2.5a the relevant scheme of a parallel arrangement of fibrils with a central core A is shown. We shall apply the term "closely-packed structure" to the structure which contains the highest possible quantity of fibrils surrounding the core A, and the corresponding "coordinational" number is denoted $Z_{\max}$. If a fibril has a perfect cylindrical shape, then obviously $Z_{\max} = 6$. If the base of the cylinder is not a circle, $Z_{\max}$ will be more than six (see

Fig. 2.5b). For the sake of brevity we shall denote the elementary cell as the f-point.

Let us estimate the density of the substance consisting of the cells shown in Fig. 2.5a. For the ideal structure of f-points the average density is

$$\rho_m^{\text{id.}} = Z_{\max} v_{fl} \rho_{fl} \frac{(1 - \mu)}{v_{\text{cell}}} \, . \tag{2.34}$$

Since

$$v_{fl} = \pi r_{fl}^2 l_{fl} \, , \quad v_{\text{cell}} = \pi R_{\text{cell}}^2 l_{fl} = 9 \pi r_{fl}^2 l_{fl} \, ,$$

where r_{fl} is the mean fibril radius, we obtain

$$\rho_m^{\text{id.}} = Z_{\text{Max}} \rho_{fl} \frac{(1 - \mu)}{9} \tag{2.35}$$

and hence at $Z_{\max} \Rightarrow 9$ the highest possible packing density is

$$\rho_{\max,m}^{\text{id.}} = \rho_{fl} (1 - \mu) < \rho_{fl} \, , \tag{2.36}$$

which is in complete accordance with the previous reasoning.

Below we shall illustrate how various parameters of such structures may be described by means of the f-points model. In particular, note in advance that the mechanical disruption force in such substances calculated by the f-points model is in good agreement with the experimental data.

2.3 Relation Between Thermodynamic Parameters and Porosity

In the investigation of the properties of porous media, it is of great practical (as well as theoretical) importance to know the dependence of thermodynamic parameters on the structure porosity m. Although we have already discussed it above, it is worth reminding ourself once more that porosity is introduced as the volume share of the pores in a substance relative to the entire dielectric volume V, that is, $m = V_p/V$.

Let us start from the Helmholtz free energy $F = F(T, V)$, where T is the temperature and V is the volume, for it is one of the most convenient thermodynamic functions from the point of view of calculations. Due to the additivity of the function F, for a two-component structure we may write

$$F = F_1 + F_2 \, , \tag{2.37}$$

where F_1 is the free energy of the main matrix, and F_2 is the free energy conditioned by the pores.

It is clear that (2.37) is the ideal presentation of the free energy of a two-component structure because it does not take into account the interaction between both phases. If we denote this interaction as, say, F_{12}, the Eq. (2.37) should be written as

$$F = F_1 + F_2 + F_{12} . \tag{2.38}$$

But in practice (to say nothing of theory), as a rule, an approximation holds when F_{12} is small compared to F_1 and F_2, and therefore the Eq. (2.37) should be chosen as the basis of analytical consideration, and this equation will be dealt with below, unless otherwise stated.

The most convenient and characteristic property of the formula for F is its linearity. Indeed, let us present it as an integral of the density of the free energy f, namely

$$F = \int\limits_V f dV = \int\limits_{V_1} f_1 dV + \int\limits_{V_2} f_2 dV , \tag{2.39}$$

where $V = V_1 + V_2$.

If f_1 in the area V_1 and f_2 in the area V_2 are homogenous functions, then taking them to be constant we can easily write (2.39) as

$$F = f_1 V_1 + f_2 V_2 . \tag{2.40}$$

Dividing (2.40) by the entire dielectric volume V and introducing the density $f = F/V$, we find

$$f = f_1(1 - m) + f_2 m . \tag{2.41}$$

The relation (2.41) is very convenient in "working" with such two-phase structures. Indeed, the entropy of a volume unit according to (2.41) and to the definition $s = \partial f / \partial T$ will be the following (compare with Eq. (3.1) of Chap. 3):

$$s = s_1(1 - m) + s_2 m . \tag{2.42}$$

The other important function is the isochoric heat capacity of a volume unit defined by the formula:

$$c_v = T \left(\frac{\partial s}{\partial T} \right)_V , \tag{2.43}$$

and hence from (2.42) it follows that

$$c_v = c_{1v}(1 - m) + c_{2v} m . \tag{2.44}$$

An analogous equation holds for the isobaric heat capacity, that is

$$c_p = c_{1p}(1 - m) + c_{2p} m . \tag{2.45}$$

The importance of the formula (2.45) is seen from the fact that c_p is included in the definition of such a significant non-equilibrium parameter as the thermal diffusion coefficient and defines its additional dependence on porosity.

Here it should be noted that for completeness, of course, the compressibility of the solid phase should be taken into account (which was done in Sect. 3.1), but as the purely illustrative approach the given formulae have the right to exist and in the overwhelming number of cases their existence is justified in practice.

The above equations do not exhaust the information on the equilibrium characteristics of porous structures because there is, for example, such a concept as the dielectric permeability ε. In this case the matter is more complicated and, in particular, Chap. 5 of this monograph is devoted to this question, but the tensor ε_{ik} is worth discussing here.

Indeed, let us introduce the definition of the tensor ε_{ik} according to the formula

$$\varepsilon_{ik} = V^{-1} \left\langle \frac{\partial^2 F}{\partial E_i^{(i)} \partial E_k^{(i)}} \right\rangle, \tag{2.46}$$

where the angular brackets mean time averaging, vector $\boldsymbol{E}^{(i)}$ is the internal structurally uniform electric field of the given heterogeneous medium, on condition that the dielectric is put into an external uniform alternating electric field $E(t) = \boldsymbol{E}_0 e^{i\omega t}$, where $\boldsymbol{E}_0$ is its amplitude, ω is the frequency, and the indices i, k characterize the three Cartesian coordinates x, y, z.

It is clear that in definition (2.46) the tensor ε_{ik} is the function of the field frequency ω, temperature T and of a number of other factors connected with the porosity and internal structure of a substance.

Here it should be noted that in spite of the seeming simplicity of (2.46), the calculation of the dependence ε_{ik} is connected with great difficulties of a purely mathematical character. Really, first of all we should remember that the internal field is a function of the fields $\boldsymbol{E}_1^{(i)}$ and $\boldsymbol{E}_2^{(i)}$ inside the corresponding phase. That is,

$$\boldsymbol{E}^{(i)} = \boldsymbol{E}^{(i)}(\boldsymbol{E}_1^{(i)}, \boldsymbol{E}_2^{(i)}). \tag{2.47}$$

If in (2.46) we substitute (2.37) and take into account the dependence (2.47), then using (2.40) we find

$$\varepsilon_{ik} = V^{-1} \left\langle \frac{\partial^2 F}{\partial E_i^{(i)} \partial E_k^{(i)}} \right\rangle = \left\langle \frac{\partial^2 f}{\partial E_i^{(i)} \partial E_k^{(i)}} \right\rangle$$

$$= (1-m) \left\langle \frac{\partial^2 f_1}{\partial E_i^{(i)} \partial E_k^{(i)}} \right\rangle + m \left\langle \frac{\partial^2 f_2}{\partial E_i^{(i)} \partial E_k^{(i)}} \right\rangle.$$

As

$$\frac{\partial f_1}{\partial E_i^{(i)}} = \left(\frac{\partial f_1}{\partial E_{1s}^{(i)}} \right) \left(\frac{\partial E_{1s}}{\partial E_i^{(i)}} \right), \quad \text{and} \quad \frac{\partial f_2}{\partial E_i^{(i)}} = \left(\frac{\partial f_2}{\partial E_{2s}^{(i)}} \right) \left(\frac{\partial E_{2s}}{\partial E_i^{(i)}} \right),$$

the second derivatives will be

$$\frac{\partial^2 f_1}{\partial E_i^{(i)} \partial E_k^{(i)}} = \left(\frac{\partial^2 f_1}{\partial E_{1s}^{(i)} \partial E_{1l}^{(i)}} \right) \left(\frac{\partial E_{1s}^{(i)}}{\partial E_i^{(i)}} \right) \left(\frac{\partial E_{1l}^{(i)}}{\partial E_i^{(i)}} \right)$$

$$+ \left(\frac{\partial f_1}{\partial E_{1s}^{(i)}} \right) \left(\frac{\partial^2 E_{1s}^{(i)}}{\partial E_i^{(i)} \partial E_k^{(i)}} \right).$$

For f_2 an analogous equation applies.

Here it is time to note that repeated indices (here and further in the text) mean summation. Introducing by definition the dielectric permeability of every particular phase

$$\begin{cases} \varepsilon_{1sl} = V^{-1} \left\langle \dfrac{\partial^2 f_1}{\partial E_{1s}^{(i)} \partial E_{1l}^{(i)}} \right\rangle , \\[2em] \varepsilon_{2sl} = V^{-1} \left\langle \dfrac{\partial^2 f_2}{\partial E_{2s}^{(i)} \partial E_{2l}^{(i)}} \right\rangle , \end{cases} \tag{2.48}$$

we obtain

$$\varepsilon_{ik} = (1-m)\varepsilon_{1sl} \left(\frac{\partial E_{1s}^{(i)}}{\partial E_i^{(i)}} \right) \left(\frac{\partial E_{1l}^{(i)}}{\partial E_i^{(i)}} \right)$$
$$+ m\varepsilon_{2sl} \left(\frac{\partial E_{2s}^{(i)}}{\partial E_i^{(i)}} \right) \left(\frac{\partial E_{2l}^{(i)}}{\partial E_i^{(i)}} \right) + B_2 , \tag{2.49}$$

where the function B_2 contains the second derivatives from the internal fields in each corresponding phase, that is

$$B_2 = \left(\frac{\partial f_1}{\partial E_{1s}^{(i)}} \right) \left(\frac{\partial^2 E_{1s}^{(i)}}{\partial E_i^{(i)} \partial E_k^{(i)}} \right) + \left(\frac{\partial f_2}{\partial E_{2s}^{(i)}} \right) \left(\frac{\partial^2 E_{2s}^{(i)}}{\partial E_i^{(i)} \partial E_k^{(i)}} \right) . \tag{2.50}$$

The general formula (2.49) may be slightly simplified if we neglect the dependence B_2, which is generally justified supposing that there exists a linear dependence between the field components $E_{1i}^{(i)}$ and $E_{2i}^{(i)}$ and the components of the resulting uniform field $E_i^{(i)}$.

Indeed, let us take derivatives:

$$\frac{\partial E_{1s}^{(i)}}{\partial E_i^{(i)}} = a_{1si} ,$$
$$\frac{\partial E_{2s}^{(i)}}{\partial E_i^{(i)}} = a_{2si} , \tag{2.51}$$

where the coefficients a_{1si} and a_{2si} are obviously some matrices whose type is not important for us now because the main thing is to understand the algorithm to calculate the physical characteristics of porous media.

Taking into account (2.51) and neglecting the function B_2 from (2.49) we have

$$\varepsilon_{ik}(\omega, T) = (1-m)\varepsilon_{1sl}(\omega, T)a_{1is}a_{1si} + m\varepsilon_{2sl}(\omega, T)a_{2is}a_{2si} . \tag{2.52}$$

The obtained formula is very convenient to estimate the frequency and temperature dependence of the dielectric permeability of a porous dielectric.

Actually, knowing tensors $\varepsilon_{1sl}(\omega, T)$ and $\varepsilon_{2sl}(\omega, T)$ it is very easy to estimate the "resulting" function $\varepsilon_{sl}(\omega, T)$ according to (2.52).

Note that if we deal with the magnetic main matrix into which the free volumes are "admixed", then analogous considerations allow us to write down the equation for the tensor of the magnetic permeability μ_{ik} using (2.52). In this case we have

$$\mu_{ik}(\omega, T) = (1 - m)\mu_{1sl}(\omega, T)b_{1is}b_{1si} + m\mu_{2sl}(\omega, T)b_{2is}b_{2si}, \qquad (2.53)$$

where $b_{1,2ik}$ are some constant matrices determined by the known rule (2.51), but which are, in general, different from the matrices $a_{1,2ik}$. For them the two following relations are correct:

$$b_{1si} = \frac{\partial H_{1s}^{(i)}}{\partial H_i^{(i)}},$$

$$b_{2si} = \frac{\partial H_{2s}^{(i)}}{\partial H_i^{(i)}}, \qquad (2.54)$$

where $H^{(i)}$ is the uniform internal magnetic field and $H_1^{(i)} H_2^{(i)}$ are the magnetic fields in the first and second phases respectively (remember that the main matrix is phase "1", and the free volumes are phase "2").

The reader may quite reasonably ask how such a non-equilibrium characteristic the dielectric permeability tensor ε_{ik} can be calculated using (2.37) for equilibrium states. The answer in fact lies on the surface. The thing is that the dependence (2.58) (as well as (2.53), of course) clears up only the simple formal connection of $\varepsilon_{ik}(\mu_{ik})$ with the porosity m! The relaxation properties themselves (that is, the non-equilibrium characteristics connected with the relaxation time τ_1 for the first phase and τ_2 for the second one) are present in tensors ε_{1ik} and ε_{2ik}, each of them having its own dependence $\tau_1(T)$ and $\tau_2(T)$. The free energy does not allow us to determine the microscopic relation between ε and τ, because for this purpose it is necessary to use a quite different mathematical apparatus and, in particular, the method of the non-equilibrium density matrix. A description of this method would go beyond the scope of this monograph, though some basic elements of the mentioned approach are present in this monograph and are described in Chap. 5.

As a simple illustration of the application of (2.52) to a real porous substance and for the sake of visualization we shall hold all tensors to be isotropic. Hence we take $\varepsilon_{ik} = \varepsilon\delta_{ik}$, $\varepsilon_{1ik} = \varepsilon_1\delta_{ik}$, $\varepsilon_{2ik} = \varepsilon_2\delta_{ik}$, where the single diagonal tensor in two dimensions is the Kronecker delta defined by the conditions $\delta_{ik} = 0$, if $i \neq k$ and $\delta_{ik} = 1$, if $i = k$. The result is

$$\varepsilon(\omega, T) = (1 - m)\varepsilon_1(\omega, T)A_1 + m\varepsilon_2(\omega, T)A_2, \qquad (2.55)$$

where new numbers introduced according to the convolution rules (the sum of diagonal elements from matrices product) are $A_1 = a_{1is}a_{1si}$, $A_2 = a_{2is}a_{2si}$.

Let us take $\varepsilon = \varepsilon' + i\varepsilon''$ and use the known formulae of Debye according to which the imaginary and real parts of the dielectric permeability coefficient are defined by the equations:

$$\varepsilon''(\omega, T) = \frac{(\varepsilon_0 - \varepsilon_\infty)\omega\tau}{1 + \omega^2\tau^2},$$

$$\varepsilon''(\omega, T) = \varepsilon_0 + \frac{(\varepsilon_0 - \varepsilon_\infty)\omega^2\tau^2}{1 + \omega^2\tau^2},$$

(2.56)

where ε_0 is the static dielectric the permeability and ε_∞ is the permeability at $\omega \to \infty$. If now we substitute (2.56) for both phases into the general relations (2.55), we shall easily find that

$$\varepsilon''(\omega, T) = (1 - m)A_1\frac{(\varepsilon_{10} - \varepsilon_{1\infty})\omega\tau_1}{1 + \omega^2\tau_1^2} + mA_2\frac{(\varepsilon_{20} - \varepsilon_{2\infty})\omega\tau_2}{1 + \omega^2\tau_2^2}, \qquad (2.57)$$

$$\varepsilon'(\omega, T) = (1 - m)A_1\left\{\varepsilon_{10} + \frac{\varepsilon_{10} - \varepsilon_{1\infty})\omega^2\tau_1^2}{1 + \omega^2\tau_1^2}\right\}$$

$$+ mA_2\varepsilon_{20}\left\{\frac{(\varepsilon_{20} - \varepsilon_{2\infty})\omega^2\tau_2^2}{1 + \omega^2\tau_2^2}\right\}. \qquad (2.58)$$

The frequency dependence is clearly seen in the given formulae. As to the temperature dependence, it "resides" in the times $\tau_1(T)$ and $\tau_2(T)$. Analysis of the obtained formulae allows one to find out quite non-standard effects connected with the absorption of alternating electric fields by porous structures (see Chap. 4).

The calculation of the times τ_1 and τ_2 should be first of all connected with the correct statement of the problem, and to give an illustrating example we shall analyze the simplest case which concerns a homogeneous (without pores!) and, moreover, crystalline dielectric. Imagine that in a sample volume the phonon subsystem is disturbed from the equilibrium condition and after this the dielectric is immediately put into a thermostat at temperature T. The task is to estimate the time necessary for all the subsystems to come to equilibrium with the thermostat.

We shall begin by isolating the main relaxing subsystems. There are three: longitudinal (l), transverse (t) and optical (o) phonons. Their interaction with each other and with the thermostat leads to internal microscopic equilibrium. After the internal equilibrium has been achieved, there begins to act the so-called Fourier law that is known to define the connection of the heat flow $\boldsymbol{q}$ with the temperature gradient ∇T. The coefficient of proportionality between $\boldsymbol{q}$ and ∇T is called the thermal conductivity coefficient and is denoted by κ. In the general case it is a tensor. The fastest process is the achievement of an internal quasi-equilibrium distribution in the system of transverse phonons. The next process is the birth of a pair of phonons from one longitudinal phonon: one transverse and one longitudinal. Due to this process a longitudinal phonon acquires its own temperature equal to the temperature of the transverse phonons T_t; the entrainment velocity $\boldsymbol{V}$ is not equal to zero and the chemical potential μ_l is not equal to zero. The return of these parameters to equilibrium values (temperature T_t tends to T, and the chemical potential and phonon entrainment velocities tend to zero) is defined by the connection with the thermostat, which is to be understood to be caused by surface phonons with an equilibrium Bose-distribution function. By taking account of such interactions we shall get linear differential equations in V, μ_l and T_t. Their solution determines the time during which these parameters tend to equilibrium. The latest relaxation time (maximum!) will thus determine the sought relation of τ with T, where $\tau = \tau_{\max}$.

For more details about methods of calculation of the non-equilibrium characteristics and, in particular, the thermal conductivity coefficient the reader is referred to Chap. 5. As to the mathematical description and derivation of the relaxation equations for the quasi-equilibrium parameters V, μ_l and T_t, they are obtained, for example, from Boltzmann's kinetic equation by its linearization over the corresponding magnitudes (V, μ_l and $\delta T = |T_t - T| \ll T$).

As an abstract example we shall describe the properties of the medium assigned in the shape of a cube with the side L, inside which balls of radius R are closely packed (without deformation). The balls will be provisionally called the main matrix and to characterize their physical properties we shall assign index "1" to the corresponding variable, and gaps between the balls which are free volumes will be characterized by the index "2".

While the total volume of the cube is $V = L^3$, and the volume of one sphere $v_1 = (4\pi/3)R^3$, then for the condition of complete packing of balls along every edge of the cube their quantity is obviously $N = (L/2R)^3$. Consequently, the porosity (the ratio of the volume between the balls to the cube volume) will be $m = 1 - Nv_1/V = 1 - (L/2R)^3(4\pi/3)(R/L)^3 = 1 - \pi/6$, and hence according to (2.44) the heat capacity turns out to be as follows:

$$c_v = \frac{\pi c_{1v}}{6} + \left(1 - \frac{\pi}{6}\right) c_{2v} \,. \tag{2.59}$$

The dielectric permeability in the isotropic approximation for the structure under consideration from (2.55) with the simplest assumption $A_1 = A_2 = 1$ becomes

$$\varepsilon = \frac{\pi \varepsilon_1}{6} + \left(1 - \frac{\pi}{6}\right) \varepsilon_2 \,. \tag{2.60}$$

The latter two very simple formulae may be used only as estimates for the description of the properties of complex heterogeneous structures. Such a step, generally speaking, is necessary not only to satisfy our purely scientific curiosity, but mainly to know for sure how the formulae for $c_v(T)$ and $\varepsilon(\omega, T)$ "work" in practice. It is always very important and useful for the very simple reason that in principle no ideal theory (or model) can be created which would give answers to all the questions of experimenters engaged in measuring the various physical properties of complex multiphase structures. To conclude this section we shall also show the connection of the main thermodynamic potentials with the corresponding independent variables as reference material.

The Gibbs potential is

$$d\Phi = -SdT + VdP\,,$$ (2.61)

the enthalpy is

$$dW = TdS + VdP\,,$$ (2.62)

and the internal energy is

$$dE = TdS - PdV\,,$$ (2.63)

where P is the pressure.

For two-phase media there are quite evident relations connected with the property of additivity of the corresponding potentials:

$$\begin{aligned} \Phi &= \Phi_1 + \Phi_2\,, \\ W &= W_1 + W_2\,, \\ E &= E_1 + E_2\,. \end{aligned}$$ (2.64)

The potentials (2.64), as related to a volume unit, may be expressed as

$$\begin{aligned} \varphi &= \frac{\Phi}{V} = (1-m)\varphi_1 + m\varphi_2\,, \\ w &= \frac{W}{V} = (1-m)w_1 + mw_2\,, \\ h &= \frac{E}{V} = (1-m)h_1 + mh_2\,. \end{aligned}$$ (2.65a)

One more formula which should be obligatorily referred to, connects via the variables P and T the Gibbs thermodynamic potential with the chemical potential denoted, as a rule, by the Greek letter μ, the latter being a function of the variables P and T. Indeed, taking $\Phi = \mu(P,T)N$, where N is the number of particles and using the additivity of the potential Φ in the case of the two-phase structure, we find that $\Phi = \mu_1 N_1 + \mu_2 N_2$, where $N_{1,2}$ is the number of particles in the corresponding phases. Dividing Φ by the total volume of the dielectric V, we find that

$$\varphi = \frac{\Phi}{V} = \mu_1 \left(\frac{N_1}{V_1}\right)\left(\frac{V_1}{V}\right) + \mu_2 \left(\frac{N_2}{V_2}\right)\left(\frac{V_2}{V}\right)$$

$$= \mu_1(P,T)(1-m)c_1 + \mu_2(P,T)mc_2\,,$$ (2.65b)

where the particle concentrations in each phase are defined by the relations $c_1 = N_1/V_1$ and $c_2 = N_2/V_2$.

Thus, knowing the chemical potentials μ_1 and μ_2, it is easy to calculate the compound action potential by the formula (2.65b).

The given formulae are of great importance in studying the physical properties both of two-phase and multiphase structures, and in the latter case they are easily obtained in the appropriate form if we denote the phase concentrations as, for example, m_i, where i is the number of different phases.

2.4 Some Data from Thermodynamics

In the process of further presentation of material we shall need a number of equations which connect some thermodynamic derivatives with others. That is why we shall now dwell upon this question in more detail and consider a number of examples enabling a prepared reader to recollect and an unprepared one to understand a simple algorithm for calculation of the corresponding relations between thermodynamic parameters of interest.

So, suppose, for example, that we are to calculate the derivative $(\partial S/\partial P)_T$, where the subscript indicates that differentiation is made at constant temperature T. As the derivative is of the entropy S with respect to the pressure P, it is clear that we deal with independent variables P and T. In these variables, as we know (see (2.61)), only the Gibbs potential Φ is written down and its total differential is $\mathrm{d}\Phi = -S\mathrm{d}T + V\mathrm{d}P$.

From this we can see that $(\partial\Phi/\partial T)_P = -S$, and $(\partial\Phi/\partial P)_T = V$. From the condition of the equality of the mixed derivatives it follows that $\partial^2\Phi/\partial T\partial P = \partial^2\Phi/\partial P\partial T$, and therefore

$$\left(\frac{\partial S}{\partial P}\right)_T = \left(\frac{\partial V}{\partial T}\right)_P . \tag{2.66}$$

Knowing the equation of the gas or liquid state from (2.66) it is easy to determine the derivative $(\partial V/\partial T)_P$. Really, if the equation of the ideal gas state (Clapeyron–Mendeleev's equation) is valid, $PV = NT$, where N is the number of gas particles (remember that we take the Boltzmann constant as equal to one, unless otherwise stated), then the derivative sought for will be $(\partial V/\partial T)_P = N/T$.

Let us now consider another and slightly more complicated example. We shall calculate the derivative $(\partial P/\partial\rho)_S$, where ρ is the density. According to the definition [10] the written derivative is nothing but the squared sound velocity in the liquid or gas. In other words, $c_s^2 = (\partial P/\partial\rho)_S$. As $\rho = M/V$, where M is the mass of liquid, then

$$c_s^2 = \left(\frac{\partial P}{\partial\rho}\right)_S = \left(\frac{\partial P}{\partial V}\right)_S \left(\frac{\partial V}{\partial\rho}\right)_S = \left(\frac{\partial P}{\partial V}\right)_S \left(\frac{\partial\rho}{\partial V}\right)_S^{-1}$$

$$= \left(\frac{\partial P}{\partial V}\right)_S \left(\frac{\partial(M/V)}{\partial V}\right)_S^{-1} = -\left(\frac{V^2}{M}\right)\left(\frac{\partial P}{\partial V}\right)_S . \tag{2.67}$$

Our task therefore is to calculate the derivative $\partial P/\partial V$ taken at constant entropy. In accordance with the above equation we deal with the variables V and S. In order to find the given derivative it is very convenient to pass from the variables V and S to the variables V and T. This procedure is rather simply performed by means of the Jacobian rule and its essence is in the following formalism. Really, if the derivative $(\partial y/\partial x)_z$ is concerned, it should be written as the Jacobian $J = \partial(y, z)/\partial(x, z)$. The transition to new variables $x - t$ is performed by the technique:

$$J = \left[\frac{\partial(y,z)}{\partial(x,z)}\right]\left[\frac{\partial(x,t)}{\partial(x,t)}\right] = \left[\frac{\partial(y,z)}{\partial(x,t)}\right]\left[\frac{\partial(x,t)}{\partial(x,z)}\right] = \left(\frac{\partial t}{\partial z}\right)_x \times \left[\frac{\partial(y,z)}{\partial(x,t)}\right].$$

Using the given rule we find

$$c_s^2 = -\left(\frac{V^2}{M}\right)\left(\frac{\partial P}{\partial V}\right)_S = -\left(\frac{V^2}{M}\right)\left[\frac{(\partial P,S)}{\partial(V,S)}\right]\left[\frac{(\partial V,T)}{\partial(V,T)}\right]$$

$$= \left(-\frac{V^2}{M}\right)\left[\frac{\partial(P,S)}{\partial(V,T)}\right]\left[\frac{\partial(V,T)}{\partial(V,S)}\right] = -\left(\frac{V^2}{M}\right)\left[\frac{\partial(P,S)}{\partial(V,T)}\right]\left[\frac{\partial(V,S)}{\partial(V,T)}\right]^{-1}.$$

But since $[\partial(V,S)/\partial(V,T)] = (\partial S/\partial T)_V = C_V/T$, it means that

$$c_s^2 = -\left(\frac{V^2 T}{MC_V}\right)\left[\frac{\partial(P,S)}{\partial(V,T)}\right] = -\begin{vmatrix} V^2T\left(\frac{\partial P}{\partial V}\right)_T & \left(\frac{\partial P}{\partial T}\right)_V \\ MC_V\left(\frac{\partial S}{\partial V}\right)_T & \left(\frac{\partial S}{\partial T}\right)_V \end{vmatrix}$$

$$= -\frac{V^2 T}{MC_V}\left[\left(\frac{\partial P}{\partial V}\right)_T\left(\frac{\partial S}{\partial T}\right)_V - \left(\frac{\partial P}{\partial T}\right)_V\left(\frac{\partial S}{\partial V}\right)_T\right]$$

$$= -\frac{V^2 T}{MC_V}\left[\left(\frac{\partial P}{\partial V}\right)_T\left(\frac{C_V}{T}\right) - \left(\frac{\partial P}{\partial T}\right)_V\left(\frac{\partial S}{\partial V}\right)_T\right]. \tag{2.68}$$

The second addend in (2.68) contains the derivative of the entropy with respect to the volume. In order to get rid of the entropy let us recollect that in the variables V and T the Helmholtz potential (free energy F) characterizes the complete differential $dF = -SdT - PdV$. Hence it follows that $(\partial F/\partial T)_V = -S$, $(\partial F/\partial V)_T = -P$. Equating the second mixed derivatives we get

$$\left(\frac{\partial S}{\partial V}\right)_T = \left(\frac{\partial P}{\partial T}\right)_V. \tag{2.69}$$

So, from (2.68) it follows that

$$c_s^2 = -\left(\frac{V^2}{M}\right)\left(\frac{\partial P}{\partial V}\right)_T + \left(\frac{V^2 T}{MC_V}\right)\left[\left(\frac{\partial P}{\partial T}\right)_V\right]^2.$$

Since (the corresponding calculations were performed above (see (2.67)) $-(V^2/M)(\partial P/\partial V)_T = (\partial P/\partial\rho)_T$, then

$$c_s^2 = \left(\frac{\partial P}{\partial\rho}\right)_T + \left(\frac{V^2 T}{MC_V}\right)\left[\left(\frac{\partial P}{\partial T}\right)_V\right]^2. \tag{2.70}$$

Let us apply the obtained formula to an ideal gas. Since the equation of state is $PV = NT$, then $(\partial P/\partial V)_T = -NT/V^2$ and $(\partial P/\partial T)_V = N/V$. Therefore, from (2.70) with taking account of what was said before Eq. (2.66), we obtain

$$c_s^2 = \frac{NT}{M} + \frac{V^2 T}{MC_V}\frac{N^2}{V^2} = \frac{NT}{M} + \frac{PVN}{MC_V} = \frac{NT}{M}\left(1 + \frac{N}{C_V}\right)$$

$$= \frac{NT}{M}\frac{C_P}{C_V} = \frac{PV}{M}\frac{C_P}{C_V} = \frac{PC_P}{\rho C_V}.$$

Here we have used the known relation, namely $C_P - C_V = N$. Thus, the sound velocity in the ideal gas is

$$c_s = \left(\frac{P\gamma}{\rho} \right)^{1/2}, \tag{2.71}$$

where the ratio of the heat capacities, C_P/C_V, is denoted by γ.

As the third and also very illustrative example of the calculation of thermodynamic derivatives, we shall find the value of the difference $C_P - C_V$ in the general case.

According to the definition [10], the isobaric heat capacity is determined as $C_P = T(\partial S/\partial T)_P$. Here we shall pass over to new variables V and T. Then we have

$$C_P = T \frac{\partial(S,P)}{\partial(T,V)} \frac{\partial(T,V)}{\partial(T,P)} = T \left(\frac{\partial V}{\partial T} \right)_T \begin{vmatrix} \left(\dfrac{\partial S}{\partial T} \right)_V & \left(\dfrac{\partial S}{\partial V} \right)_T \\ \left(\dfrac{\partial P}{\partial T} \right)_V & \left(\dfrac{\partial P}{\partial V} \right)_T \end{vmatrix}$$

$$= T \left(\frac{\partial V}{\partial T} \right)_T \left[\left(\frac{\partial S}{\partial T} \right)_V \left(\frac{\partial P}{\partial V} \right)_T - \left(\frac{\partial S}{\partial V} \right)_T \left(\frac{\partial P}{\partial T} \right)_V \right].$$

Removing the brackets we find

$$C_P = T \left(\frac{\partial S}{\partial T} \right)_V - T \left(\frac{\partial V}{\partial P} \right)_T \left(\frac{\partial P}{\partial T} \right)_V \left(\frac{\partial S}{\partial V} \right)_T.$$

According to (2.69), $(\partial S/\partial V)_T = (\partial P/\partial T)_V$ and $T(\partial S/\partial T)_V = C_V$, that is why for the sought difference we get

$$C_P - C_V = T \left(\frac{\partial V}{\partial P} \right)_T \left[\left(\frac{\partial P}{\partial T} \right)_V \right]^2. \tag{2.72}$$

If we use the equation of state $PV = NT$, we shall finally find the formula which we already know, that is, $C_P - C_V = N$. The reader is recommended to verify the validity of the given relation independently. Note, by the way, that in the following section we shall need the formula (2.72).

2.5 Porous Media as very Viscous Liquids with Bubbles

If we divert for a moment our attention from complex and highly non-ordered porous media considered in the present monograph, such as cellulose and paper, and imagine, for example, just a liquid with inclusions in the form of air (gas) bubbles with a relative volume concentration m, the estimation of the dependence of $c_s(m)$ for such two-phase substances shows a rather specific behavior of the sound velocity as a function of m. Let us present it mathematically. If $m = V_2/V$, where V_2 is the volume of the bubbles, and

$V = V_1 + V_2$, where V_1 is the liquid volume, then the entropy related to a volume unit will be

$$s = \frac{S}{V} = \left(\frac{S_1}{V_1}\right)\left(\frac{V_1}{V}\right) + \left(\frac{S_2}{V_2}\right)\left(\frac{V_2}{V}\right) = s_1(1 - m) + s_2 m\,. \tag{2.73}$$

The density will be introduced in a similiar way:

$$\rho = \rho_1(1 - m) + \rho_2 m\,. \tag{2.74}$$

We shall further consider that the following two conditions are satisfied:

1. All the substance under consideration is in a state of thermodynamic equilibrium. It means that the change of the pressure P with the temperature T depends on the Clapeyron–Clausius equation:

$$\frac{\mathrm{d}P}{\mathrm{d}T} = \frac{(s_2 - s_1)}{(v_2 - v_1)}\,, \tag{2.75}$$

   where s_1 and s_2 are the entropy, and v_1 and v_2 are the volumes, related to the particles of the corresponding phases.

2. The sound wavelength λ considerably exceeds the linear dimensions of the inhomogeneities (bubble diameters), but at the same time the wavelength is considerably smaller than the distance between the bubbles, d. In short, $2R \ll \lambda \ll \mathrm{d}$, where R is the bubble radius. The sound velocity in a gas is known to be determined by the equation:

$$c_s = \left(\frac{\partial P}{\partial \rho}\right)_S\,, \tag{2.76}$$

Therefore, if we pass from variables S, ρ to more convenient variables ρ, m, we obtain

$$\left(\frac{\partial P}{\partial \rho}\right)_S = \frac{\partial(P, S)}{\partial(\rho, S)}\frac{\partial(\rho, m)}{\partial(\rho, m)} = \frac{\partial(P, S)}{\partial(\rho, m)}\frac{\partial(\rho, m)}{\partial(\rho, S)}$$

$$= \left(\frac{\partial S}{\partial m}\right)_\rho^{-1} \begin{vmatrix} \left(\dfrac{\partial P}{\partial \rho}\right)_m & \left(\dfrac{\partial P}{\partial m}\right)_\rho \\[2ex] \left(\dfrac{\partial S}{\partial \rho}\right)_m & \left(\dfrac{\partial S}{\partial m}\right)_\rho \end{vmatrix}\,. \tag{2.77}$$

If we now calculate the determinant, we obtain

$$\left(\frac{\partial P}{\partial \rho}\right)_S = \left(\frac{\partial P}{\partial \rho}\right)_m - \left(\frac{\partial P}{\partial S}\right)_\rho\left(\frac{\partial S}{\partial \rho}\right)_m\,. \tag{2.78}$$

Since the density $\rho = M/V$, and M is a fixed gas mass, actually the corresponding derivative in (2.77) must be taken at constant volume V. Indeed, according to (2.73)

$$\left(\frac{\partial S}{\partial m}\right)_\rho = \left(\frac{\partial S}{\partial m}\right)_V$$

$$= V(s_2 - s_1) + V(1 - m)\left(\frac{\partial S_1}{\partial m}\right)_V + Vm\left(\frac{\partial S_2}{\partial m}\right)_V . \quad (2.79)$$

Then from (2.73) it also follows that

$$\left(\frac{\partial \rho}{\partial P}\right)_m = m\left(\frac{\partial \rho_2}{\partial P}\right)_m + (1 - m)\left(\frac{\partial \rho_1}{\partial P}\right)_m . \quad (2.80)$$

Let us calculate the derivatives contained here. We have

$$\left(\frac{\partial \rho_2}{\partial P}\right)_m = M_2\left[\left(\frac{\partial}{\partial P}\right)\left(\frac{1}{V_2}\right)\right]_m$$

$$= -\left(\frac{M_2}{V_2^2}\right)\left(\frac{\partial V_2}{\partial P}\right)_m\left(\frac{\mathrm{d}T}{\mathrm{d}P}\right) . \quad (2.81)$$

But the derivative $(\partial V_2/\partial P)_m = -\alpha_2 V_2$, where α_2 is the thermal expansion coefficient of the bubbles. As a result, from (2.80) we get

$$\left(\frac{\partial \rho_2}{\partial P}\right)_m = \rho_2\alpha_2\left(\frac{\mathrm{d}T}{\mathrm{d}P}\right) . \quad (2.82)$$

By analogy we get

$$\left(\frac{\partial \rho_1}{\partial P}\right)_m = \rho_1\alpha_1\left(\frac{\mathrm{d}T}{\mathrm{d}P}\right) . \quad (2.83)$$

Consequently, Eq. (2.79), taking account of the equalities (2.81) and (2.82), gives

$$\left(\frac{\partial \rho}{\partial P}\right)_m = [\alpha_1\rho_1(1 - m) + \alpha_2\rho_2 m]\left(\frac{\mathrm{d}T}{\mathrm{d}P}\right) .$$

3. By means of the Clapeyron–Clausius equation (2.75) we finally find

$$\left(\frac{\partial \rho}{\partial P}\right)_m = [\alpha_1\rho_1(1 - m) + \alpha_2\rho_2 m]\left[\frac{(v_2 - v_1)}{(s_2 - s_1)}\right] . \quad (2.84)$$

As to the derivative $(\partial S/\partial \rho)_m$ from (2.77), according to (2.73) we get

$$\left(\frac{\partial S}{\partial \rho}\right)_m = m\left(\frac{\partial S_2}{\partial \rho}\right)_m + (1 - m)\left(\frac{\partial S_1}{\partial \rho}\right)_m . \quad (2.85)$$

Then,

$$\left(\frac{\partial S_2}{\partial \rho}\right)_m = \left(\frac{\partial \rho}{\partial S_2}\right)_m^{-1} = \frac{1}{(\partial/\partial S_2)[(1 - m)\rho_1 + m\rho_2]}$$

$$= m^{-1}\left(\frac{\partial S_2}{\partial \rho_2}\right) = -\left(\frac{V_2^2}{M_2 m}\right)\left(\frac{\partial S_2}{\partial V_2}\right)_m .$$

Here we have taken into account that due to the independence of phase "1" properties from phase "2" properties, we may take $(\partial \rho_1 / \partial S_2)_m = 0$. We have

$$\left(\frac{\partial S_1}{\partial \rho}\right)_m = \left(\frac{\partial \rho}{\partial S_1}\right)_m^{-1}$$

$$= (1-m)^{-1} \left(\frac{\partial S_1}{\partial \rho_1}\right) = - \left(\frac{V_1^2}{M_1}\right) (1-m) \left(\frac{\partial S_1}{\partial V_1}\right)_m .$$

And, consequently, with the account of the found derivatives (2.84) gives

$$\left(\frac{\partial S}{\partial \rho}\right)_m = - \left[\left(\frac{V_2^2}{M_2}\right)\left(\frac{\partial S_2}{\partial V_2}\right)_m + \left(\frac{V_1^2}{M_1}\right)\left(\frac{\partial S_1}{\partial V_1}\right)_m\right]. \qquad (2.86)$$

According to the formula known from thermodynamics the total differential of the Helmholtz free energy is $dF = -SdT - PdV$. From the condition of the equality of mixed derivatives it follows that $(\partial S / \partial V)_T = (\partial P / \partial T)_V$. In conditions of thermodynamic equilibrium for $V = \text{const}$ the equation $(\partial P / \partial T)_V = dP/dT$ is valid. This, in turn, means that

$$\left(\frac{\partial S_1}{\partial V_1}\right)_m = \left(\frac{\partial S_2}{\partial V_2}\right)_m = \frac{dP}{dT}. \qquad (2.87)$$

Therefore, by means of (2.76) from (2.85) we find

$$\left(\frac{\partial S}{\partial \rho}\right)_m = - \left[\left(\frac{V_2^2}{M_2}\right) + \left(\frac{V_1^2}{M_1}\right)\right]\left(\frac{dP}{dT}\right)$$

$$= - \left(\frac{M_1}{\rho_1^2} + \frac{M_2}{\rho_2^2}\right)\left(\frac{dP}{dT}\right) = -M \left[\frac{(1-m)}{\rho_1^2} + \frac{m}{\rho_2^2}\right]\left(\frac{dP}{dT}\right)$$

and, due to the Clapeyron–Clausius equation (2.75), we finally have

$$\left(\frac{\partial S}{\partial \rho}\right)_m = -M \left[\frac{(1-m)}{\rho_1^2} + \frac{m}{\rho_2^2}\right]\left[\frac{(s_2 - s_1)}{v_2 - v_1}\right]. \qquad (2.88)$$

Thus, in (2.77) we have to estimate only the derivative $(\partial P / \partial S)_\rho$. Indeed,

$$\left(\frac{\partial P}{\partial S}\right)_\rho = \left(\frac{\partial P}{\partial S}\right)_V = \left(\frac{\partial P}{\partial T}\right)_V \left(\frac{\partial T}{\partial S}\right)_V = \left(\frac{dP}{dT}\right)\left(\frac{T}{C_V}\right). \qquad (2.89)$$

The isochoric heat capacity is introduced in a "conventional" way: $C_V = T(\partial S / \partial T)_V$. Taking account of (2.45) we have

$$C_V = V \left[(1-m)c_{1V} + mc_{2V}\right]. \qquad (2.90)$$

Thus, the sought-after dependence of the sound velocity on the vapour bubble concentration m, according to (2.80), into which the formulae (2.75), (2.84) and (2.88)–(2.90) should be substituted, is

$$c_s^2(m) = \frac{s_1 - s_2}{(v_2 - v_1)[m\alpha_2\rho_2 + (1-m)\alpha_1\rho_1]}$$

$$+ \frac{TM(s_2 - s_1)^2}{C_V(v_2 - v_1)^2}\left(\frac{m}{\rho_2^2} + \frac{(1-m)}{\rho_1^2}\right). \tag{2.91}$$

Let us now examine the Eq. (2.90) in the limiting case when m tends to zero. Indeed, taking $m = 0$ and omitting index "1" in all the magnitudes and holding $(s_2 - s_1)/(v_2 - v_1) = (\partial S/\partial V)_T$, we find that

$$c_s^2 = -\frac{1}{\alpha\rho}\left(\frac{\partial S}{\partial V}\right)_T + \left(\frac{\partial P}{\partial T}\right)_V \frac{TM}{C_V\rho^2}.$$

As $\alpha = -V^{-1}(\partial V/\partial T)_P$ and the density $\rho = M/V$, it directly follows that

$$c_s^2 = -\frac{V^2}{M}\left(\frac{\partial S}{\partial V}\right)_T \frac{1}{(\partial V/\partial T)_P} + \left(\frac{\partial P}{\partial T}\right)_V^2 \frac{TV^2}{C_V M}.$$

Using the known relations between the thermodynamic derivatives, namely

$$\left(\frac{\partial S}{\partial V}\right)_T = \left(\frac{\partial P}{\partial T}\right)_V, \quad \left(\frac{\partial V}{\partial T}\right)_P = -\left(\frac{\partial P}{\partial T}\right)_V\left(\frac{\partial V}{\partial P}\right)_T,$$

we get for the first addend

$$\frac{V^2}{M}\left(\frac{\partial S}{\partial V}\right)_T \frac{1}{(\partial V/\partial T)_P} = -\left(\frac{\partial P}{\partial T}\right)_V \frac{V^2}{M} \frac{1}{(\partial P/\partial T)_V(\partial V/\partial P)_T}$$

$$= -\frac{V^2}{M}\left(\frac{\partial P}{\partial V}\right)_T = \left(\frac{\partial P}{\partial \rho}\right)_T.$$

Therefore,

$$c_s^2 = \left(\frac{\partial P}{\partial \rho}\right)_T + \left(\frac{TV^2}{MC_V}\right)\left(\frac{\partial P}{\partial T}\right)_V^2. \tag{2.92}$$

In order to prove the Eq. (2.91) we should remember that for an ideal liquid the squared sound velocity is

$$c_s^2 = \left(\frac{C_P}{C_V}\right)\left(\frac{\partial P}{\partial \rho}\right)_T, \tag{2.93}$$

where C_P is the isobaric heat capacity. Equating (2.91) and (2.92), we find

$$-\left(\frac{V^2}{M}\right)\left(\frac{C_P}{C_V}\right)\left(\frac{\partial P}{\partial V}\right)_T = -\left(\frac{V^2}{M}\right)\left(\frac{\partial P}{\partial V}\right)_T + \left(\frac{TV^2}{MC_V}\right)\left(\frac{\partial P}{\partial T}\right)_V^2.$$

As the result of simple algebraic transformations (which we recommend to be performed independently), it follows that the equation

$$C_P - C_V = -T\frac{\left[\left(\frac{\partial P}{\partial T}\right)_V\right]^2}{\left(\frac{\partial P}{\partial V}\right)_T}$$

should be valid. Again, according to the known thermodynamic relation (see (2.72)) we are convinced that we have got identity. Thus we have proved that in the limiting case when the bubble concentration tends to zero, the Eq. (2.90) transforms into a correct thermodynamic formula. Note, by the way, that for $m \rightarrow 1$ Eq. (2.90) will lead to the sound velocity in the gas phase, i.e. also to a correct relation.

Finally, we should note that, to determine the dependence of the sound velocity in a porous structure from the gas phase concentration, in Sect. 3.1 we shall adhere to the more convenient, as it seems to us, concept connected with finding the derivative not of $(\partial P/\partial \rho)_S$ but the inverse derivative $(\partial \rho/\partial P)_S$. It is clear that both approaches are equivalent. However, as we shall see, in the latter case some curious peculiarities may be found, which are "lost" in the example considered above.

3 Equilibrium Physical Parameters of Porous Dielectrics

In this chapter we shall consider the main equilibrium physical parameters characterizing highly disordered dielectrics and give a rigorous theoretical interpretation based on comparison with experiment.

Correct introduction of purely physical terminology for practical use will make scientifically strict the production processes which are used to obtain porous substances. That is why it is worthwhile beginning to discuss the physical characteristics of a strictly equilibrium system. As the present monograph is also intended for technical workers in general, to begin with it would be appropriate to give the definition of what we imply by the term equilibrium system.

So, a system is called an equilibrium system if after a long period of time t_0 all the phenomena connected with the release and absorption of energy (so-called dissipation phenomena) have stopped changing. If, for example, in a dielectric the equilibrium temperature has reached a steady value equal to the temperature of the environment, it means that the process of heat conduction has stopped and a characteristic time has passed equal in order of magnitude to $t_0 \cong L^2/\chi$, where L is the characteristic linear dimension of the sample and χ is its thermal diffusion.

The present chapter deals with systems after the time systems when all the micro- and macroscopic processes for achieving internal equilibrium have been completed.

3.1 The Sound Velocity in Porous Substances

The main characteristic of substances containing voids (free volumes) is, as we know, structure porosity determined by the relation $m = V_p/V^1$, where V_p is the volume of the pores and V is the volume of the sample. It is clear that the value m is the parameter which characterizes such technically important values as the electric breakdown field, thermal conductivity, mechanical strength, density and so on.

---

[1] Sometimes the porous structure is understood as the ratio of the pore volume to the volume occupied by the substance itself, i.e. $m^* = V_p/V_f$ and denoted by l. We shall confine ourselves to the characteristic m. As to their relation, it is obvious: $l = m/(1-m)$.

It is interesting to note that in spite of the importance of the index of "void" of the m structure its determination through some parameters of the sample (equilibrium or non-equilibrium) so far not received attention. For this reason the experimental determination of m has some complications connected with the lack of information on the parametrical dependence of this value. Further, we shall give the relevant formulae allowing us to bridge this gap.

The easiest and the most effective method to determine the porosity is, in our opinion, to measure the dependence of the sound velocity c_s in a sample at frequencies corresponding to the wavelength $\lambda \gg b_{\max}$, where $b_{\max}$ is the maximum dimension of the structure inhomogeneity (the dimension of the pores!).

Let us now, by convention, divide the whole volume of the dielectric into two phases. We assign the index "1" to all the parameters characterizing the fiber (fibril) structure and the index "2" to the free volumes.

Then the entropy of the entire dielectric will be

$$S = (1 - m)S_1 + mS_2 \tag{3.1}$$

and the density

$$\rho = (1 - m)\rho_1 + m\rho_2 . \tag{3.2}$$

In order to calculate the dependence $c_s(m)$ we shall use the formula [10]:

$$c_s^2 = \left(\frac{\partial P}{\partial \rho}\right)_S , \tag{3.3}$$

where P is some conventional pressure.

The variables P and T are the most convenient in the framework of our task. The Eq. (3.3) in these variables will be

$$c_s^2 = \left[\frac{\partial(P,S)}{\partial(\rho,S)}\right]\left[\frac{\partial(P,T)}{\partial(P,T)}\right]$$

$$= \frac{C_P}{T[(\partial\rho/\partial P)_T(\partial S/\partial T)_P - (\partial S/\partial P)_T(\partial\rho/\partial T)_P]} , \tag{3.4}$$

where $C_p = (1 - m)C_{1P} + mC_{1P}$, C_{1P} is heat capacity of the solid phase at constant pressure, and C_{2P} is the heat capacity of the gas filling the free volumes.

Substituting (3.1) and (3.2) into (3.4) we find

$$c_s^{-2} = -\left(\frac{T}{C_P}\right)\left\{\left(\frac{C_P}{T}\right)\left[(1 - m)\left(\frac{\partial V_1}{\partial P}\right)_T\frac{\rho_1}{V_1} + m\left(\frac{\partial V_2}{\partial P}\right)_T\frac{\rho_1}{V_2}\right.\right.$$

$$\left.\left. + \left(\frac{\partial V_2}{\partial P}\right)_T\frac{(S_1 - S_2)}{V} + (1 - m)\left(\frac{\partial V_1}{\partial P}\right)_T + m\left(\frac{\partial V_2}{\partial T}\right)_P\right]\right.$$

$$\times \left[m \left(\frac{\partial V_2}{\partial T} \right)_P \frac{\rho_1}{V} + (1 - m) \left(\frac{\partial V_1}{\partial T} \right)_P \frac{\rho_1}{V_1} \right] \right\} . \tag{3.5}$$

In order to calculate the present derivatives, we shall consider that the gas filling the pores is ideal. Such an assumption is very often justified in practice, for the gas phase is a rarefied medium and hence the interaction between the gas molecules is weak. For this reason the equation of Clapeyron–Mendeleev is valid:

$$PV_2 = N_2 T , \tag{3.6}$$

where N_2 is the number of molecules in the volume V_2. Further, unless otherwise stated, we shall use the power system of units, that is we take the Boltzmann constant $k_B = 1$. Let us introduce two compressibilities of the solid phase: isothermal,

$$\beta_T = -V_1^{-1} \left(\frac{\partial V_1}{\partial P} \right)_T , \tag{3.7}$$

and isobaric,

$$\beta_P = V_1^{-1} \left(\frac{\partial V_1}{\partial T} \right)_P . \tag{3.8}$$

Then, taking account of (3.6), (3.7) and (3.8) we find

$$c_s^{-2} = \rho_1 \left[(1 - m)\beta_T + \frac{m}{P} \right] - \rho_1 V \left[m + (1 - m)T\beta_P \right]$$

$$\times \frac{[m^2 + (1 - m)^2 T\beta_T + mq/PV]}{TC_P} , \tag{3.9}$$

where

$$q = T(S_2 - S_1) , \quad C_P = (1 - m)[C_{1V} + (1 - m)\beta_P T] + mC_{2P} . \tag{3.10}$$

Supposing that the pressure in the gas phase weakly affects the volume of the solid phase structure (fiber) we may approximately take $\beta_T \cong 0$. Besides, we shall also suppose that the temperature rather weakly affects the fiber volume, i.e. $\beta_P \cong 0$. In this case we obtain the following simple equation for the inverse sound velocity:

$$c_s^{-2} = \left(\frac{m\rho_1}{P} \right) \left\{ 1 - m \frac{mPV + q}{[mC_{2P} + (1 - m)C_{1V}]T} \right\}$$

$$+ (1 - m)c_{sf}^{-2} + mc_{sp}^{-2} . \tag{3.11}$$

The latter two addends in (3.11) are introduced only in order to get "correct" equations for the square of the inverse sound velocity in the two limiting cases: $m \Rightarrow 0$ and $m \Rightarrow 1$. In the intermediate case, which will be dealt with, they should be omitted.

The formula (3.11) may be transformed into a more convenient form if the equation of state $PV_2/T = N_2$ is used. Indeed,

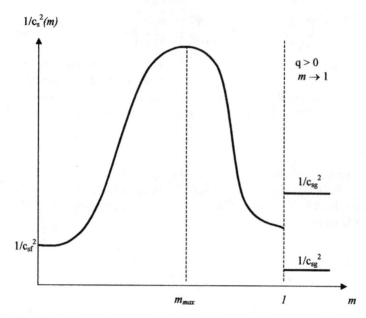

Fig. 3.1. Schematic representation of the sound velocity dependence on the porosity (m) at $q > 0$ and parameter m near one $(m \Rightarrow 1)$. C_{sg} is sound velocity in the gaseous phase and c_{sfl} is the sound velocity in the fiber.

$$c_s^2(m) = \frac{P[mC_{2P} + (1-m)C_{1V}]}{m\rho_1[mC_{2V} + (1-m)C_{1V} + m(S_1 - S_2)]} , \qquad (3.12)$$

where $C_{2P} = C_{2V} + N_2$.

The plot of the function $c_s^2(m)$ is shown in Fig. 3.1. The equation obtained includes all the known macroscopic parameters of the substance under investigation. Indeed, the heat capacities C_{1V} and C_{2V} are found experimentally (remember that C_{1V} is the isochoric heat capacity of the pure matrix in the absence of free volumes; C_{2V} is the isochoric heat capacity of the gas in the free volumes (pores). If we deal with a Maxwell gas, then $C_{2V} = $ const. and the entropies of both phases are readily found based on the known isochoric heat capacities. Indeed,

$$S_1 = N_1 \left\{ \ln\left(\frac{eV_1}{N_1}\right) + \xi_1 + \left(\frac{\partial}{\partial T}\right) \left[\int\limits_0^T \left(\int\limits_0^T C_{1V}(T')\frac{\mathrm{d}T'}{T'} \right) \mathrm{d}T \right] \right\} \qquad (3.13)$$

and analogously for S_2. So the heat flow may be determined as $q = T(S_2 - S_1)$. As far as the pressure P is concerned, it is known from the technology for producing the given structure. All the rest parameters present in (3.12) may also be considered as predetermined.

It should be noted that from the derived Eq. (3.12) a very interesting conclusion may be derived. Looking attentively at Fig. 3.1 one can't help paying attention to the fact that the sound velocity in a porous medium in a definite range of m values turns out to be less than in the gas phase! This phenomenon may be explained in the following way.

At certain porosity m in the range

$$0 < m < \frac{q}{2PV} + \left[\left(\frac{q}{4PV}\right)^2 + \frac{T}{PV}\right]^{1/2} \tag{3.14}$$

the function $c_s^{-2}(m)$ characterized by three independent parameters, temperature, pressure in the gas phase and volume, will first rise and then drop. If it turns out that $q > 0$ (for some class of substances), the peak of the curve will be shifted to the point $m = 1$ (solid line in Fig. 3.1). If $q < 0$, the peak is closer to the point $m = 0$ (dotted line). It is clear that the given effect (decrease of the sound velocity in a porous medium as compared to in a purely gaseous phase) is accounted for only by the presence of two phases and is completely determined by the heat of the "transition", q (the entropy difference of both phases!). It is as if each medium "senses" the other one and the exchange of information between them takes place in such a way as to observe the principle of Le Chatelier, according to which every medium tends to reduce the influence of the other. If there is only one phase (either fiber or gas), the sound velocity is a simple continuous function of the corresponding medium parameters.

And finally, if a dielectric is highly porous ($m \Rightarrow 1$) and $q < 0$, then the behavior of the function $c_s^{-2}(m)$ visualized in Fig. 3.2 becomes qualitatively different.

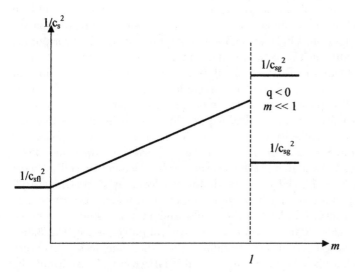

Fig. 3.2. Schematic dependence of the sound velocity in the case of a negative heat of transition ($q < 0$) and at small values of m

Table 3.1. Sound velocity in various types of porous structures at room temperature

N	Type of cellulose	c_s (m/s)
1	Viscose (coniferous) BAIC (Bratsk)	182
2	Viscose (spruce) (Kotlas)	125.4
3	E-2 MASHN	178.5
4	Sulphate bleached (coniferous) viscose (Ust-Ilimsk)	138
5	Capacitor paper	200
6	Cardicell-supercord (coniferous) (Sweden)	175

As to the experiment which was conducted to measure the sound velocity in different cellulose samples of home and foreign production, it is in complete agreement with the above theory and proves that highly porous substances may be considered as two-phase systems in which the sound velocity is less than in gas medium for m of order of one. The given statement is illustrated in Table 3.1, where the numerical values of the sound velocity are given, measured at room temperature.

3.2 The Heat Capacity of Dielectrics with Ideal Packing of Fibrils

In considering the question of the solid-phase heat capacity, Einstein's model treating the ensemble of atoms as a set of linear oscillators should be helpful [12]. Such a model allows us to describe correctly the heat capacity $C(T)$ of crystal structures both at high and low temperatures. But for porous dielectrics, typically such as cellulose and paper, the application of this model appears problematic. First of all, it is caused by the fact that in real fiber material the role of the solid phase is played by the fiber walls consisting of fibrils. The part occupied by remaining free volumes proves to be temperature-insensitive. So, if we deal with a "fibrous" dielectric, we may try to calculate the heat capacity $C(T)$ within the framework of the "f-points" model described in Sect. 2.2. In order to make the corresponding calculations let us turn to Sect. 2.2 and define the structure of the "elementary cell" as a set of f-points (Fig. 3.3). The latter figure shows that formally the "elementary" structure consisting of orderly arranged f-points is a macro-prototype of a crystal atomic structure. Indeed, if the fibrils are laid in the shape of a cube, then on the atomic scale it corresponds to a crystal cubic lattice (Fig. 3.3a). If we insert one more fibril into the middle of the cube, then such a structure will be very much a like a volume-centered cubic lattice (Fig. 3.3b). Similarly, other types of elementary structure may be packed. In this case the role of the exchange interaction, unlike in crystals, is played by the Van der Waals interaction.

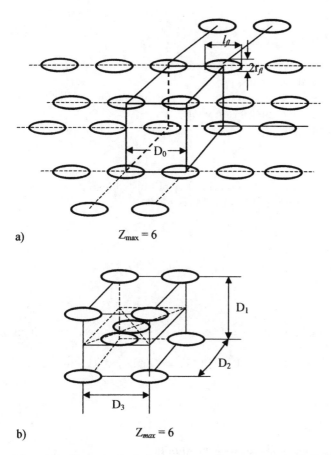

Fig. 3.3. Hypothetical elementary cell constructed of fibril links in the form of periodically repeating in directions of axes x, y, z cubes with "step" D_0 (**a**) and volume-centered fibril structure (**b**). Periods on axes x, y, z are equal to D_1 , D_2, D_3, respectively.

It should be stressed once more that in studying some physical parameters of the dielectric under investigation such a type of real hypothetical structure should be modeled which would be able to interpret the largest number of experiments as accurately as possible.

In order to determine the temperature dependence of the heat capacity of the substance whose structure is illustrated in Fig. 3.3a, we shall use the apparatus of statistical physics [11]. For this purpose we shall write the equation for the free energy as

$$F = -T \ln Z , \tag{3.15}$$

where the statistical sum

$$Z = \sum_{\{n\}} \exp\left(-\frac{E_n}{T}\right) \tag{3.16}$$

and the summing is done over all the degrees of freedom and dynamic characteristics of the internal structure.

In order to calculate Z it is necessary to find the dependence of the energy E_n. For Fig. 3.3a the average interaction energy is

$$U = \sum_{i,j}^{Z_{\max}} U_{ij}, \tag{3.17}$$

where the energy of the pairwise interaction U_{ij} of fibrils located in "sites" i and j according to monographs [13] and [14] is

$$U_{ij} = -\left(\frac{\pi^2 \hbar c \sigma_{fl} N_{fl}}{720|r_i - r_j|^3}\right)\left[\frac{(\varepsilon - 1)}{(\varepsilon + 2)}\right]^2, \tag{3.18}$$

where ε is the dielectric permeability of the fibrils, and σ_{fl} is the half-square of the cylinder fibril surface

$$\sigma_{fl} = \pi r_{fl} l_{fl}, \tag{3.19}$$

where $\hbar$ is Planck's constant and $c = 3 \times 10^{10}$ cm/s is the speed of light in vacuum.

Substituting (3.18) with (3.19) into (3.17) we find that the interaction energy is

$$U = -\left(\frac{\pi^3 Z_{\max} \hbar c l_{fl} N_{fl}}{2880 r_{fl}^2}\right)\left[\frac{(\varepsilon - 1)}{(\varepsilon + 2)}\right]^2, \tag{3.20}$$

where $Z_{\max}$ is the number of the nearest neighbors.

For example, for the structure shown in Fig. 3.3a, $Z_{\max} = 6$, and for Fig. 3.3b the characteristic value of the coordinational number $Z_{\max} = 8$.

Separating the energy E_n between the interatomic and the intermolecular interaction ε_n inside the fibrils, we express it as

$$E_n = \varepsilon_n + U. \tag{3.21}$$

Substituting (3.21) into (3.16) and then into (3.15), we get for the free energy:

$$F = \Delta F + U, \tag{3.22}$$

where

$$\Delta F = -T \ln \sum_{\{n\}} \exp\{-\varepsilon_n/T\}.$$

As the isochoric heat capacity

$$C_V = T\left(\frac{\partial S}{\partial T}\right)_V = -T\left(\frac{\partial^2 F}{\partial T^2}\right)_V, \tag{3.23}$$

where $V = V_p + V_{fl}$, then differentiating doubly Eq. (3.22) with respect to temperature we get

$$C_V = \frac{\Delta C_V}{N_{fl}} + \Phi(T), \tag{3.24}$$

where the constant component of heat capacity is given by

$$\Delta C_V = -T \left(\frac{\partial^2 F}{\partial T^2} \right)_V,$$

and the function $\Phi(T)$ is given by

$$\Phi(T) = -\frac{T \partial^2 U}{\partial T^2} = \left(\frac{\pi^3 Z_{max} T \hbar c l_{fl}}{720 r_{fl}^3} \right)$$

$$\times \left\{ 2 \left[\frac{(2-\varepsilon)}{(2+\varepsilon)^4} \right] \left(\frac{\partial \varepsilon}{\partial T} \right)_V^2 + \left[\frac{(\varepsilon-1)}{(\varepsilon+2)} \right]^3 \left(\frac{\partial^2 \varepsilon}{\partial T^2} \right)_V \right\}. \tag{3.25}$$

From the obtained equation we may make a very important conclusion that in order to describe adequately the heat capacity of the porous structure, it is necessary to know the dependence of the dielectric permeability on T only for the fibrils. Their structure is very close to a crystal one. In this respect the fact becomes very obvious that for a porous medium ε is a function with a very weak temperature dependence, and therefore the heat capacity of such media changes also very slowly (Fig. 3.4). It should be stressed that this

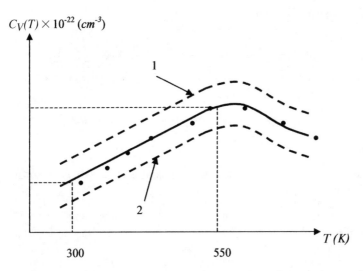

Fig. 3.4. Temperature dependence of the isochoric heat capacity of a porous substance the solid line is theory, and the dotted lines 1 and 2 describe, respectively, the cases when products $K_1 K_2 < 1$ and $K_1 K_2 > 1$. The points in the figure are the experimental points [63].

statement refers to the range of room temperatures. The circles in Fig. 3.4 correspond to the experiments discussed in Sect. 3.4.

3.3 The Transition from an Ideal Substance to a Real One

In the present section we shall deal with real rather than abstract models and the main purpose of this section is to give general methods of transition from the ideal internal structure of porous dielectrics to the "natural" one. In order to design some ways in this direction we could turn to the photographs obtained by electron microscopy. But in principle there is no such necessity so far, because it is quite obvious that the ideal cylindrical structure of fibers does not exist. Thus the idea suggests itself of the introduction of some correction coefficient, which we shall denote K_1, taking into account non-cylindrical shape of the fiber structure.

Besides, in order to take into account the imperfect (chaotic) arrangement of fibers in the dielectric volume we shall introduce the coefficient K_2. Using these corrections, the volume occupied by the real fibers may be expressed as follows:

$$V_f = N_f v_{id} K_1 K_2 \,, \tag{3.26}$$

and therefore the real porosity is

Table 3.2. Heat capacity of cellulose, $C_V \cdot 10^{22}$ (I/g)

N	Type of cellulose	Temperature T (°C)										
		25	50	75	100	125	150	175	200	225	250	275
1	Viscose sulphate low-viscous coniferous (Baikal)	8.75	4.5	10.0	14.8	6.87	–	–	4.75	7.24	7.4	7.0
2	Viscose leaf-bearing BTIC (Bratsk timber-industrial complex)	8.25	9.76	7.52	15.5	12.4	10.4	2.84	5.15	6.47	17.9	10.2
3	Viscose sulphate bleached coniferous (Ust-Ilimsk)	14.9	10.7	11.3	14.6	16.9	9.53	16.2	11.1	6.65	6.22	5.9
4	E-2 MASHN	12.9	8.94	16.5	16.1	10.3	5.17	7.44	8.01	5.74	5.37	4.4
5	Sulphate coniferous 979 (Baikal)	8.74	8.64	18.9	19.5	12.1	7.17	–	10.6	10.4	–	11.3
6	Tyredell (USA)	13.1	9.7	11.2	22.5	16.0	18.2	11.2	9.74	7.87	10.4	10.0
7	Viscose coniferous (Kotlas)	8.64	13.6	9.12	21.8	7.9	15.4	12.2	16.6	6.9	6.4	5.9
8	Finnish viscose (Raushrepala)	4.07	5.3	11.4	17.0	13.3	15.7	7.5	10.5	9.0	3.74	3.43

$$m_r = 1 - \frac{V_f}{V} = 1 - (1 - m)K_1 K_2 \,, \tag{3.27}$$

where

$$1 - m = \frac{N_f v_{id}}{V} \,.$$

Thus, the volume occupied by a real fibril is

$$v_{fl}^r = \frac{V_f}{N_f N_{fl}} = \frac{K_1 K_2 V_{id}}{N_{fl}} = K_1 K_2 v_{fl}^{id} \,, \tag{3.28}$$

where v_{fl}^{id} is the volume of an ideal fibril.

As, on the other hand,

$$v_{fl}^r = (r_{fl}^r)^2 l_{fl}^r = K_1 K_2 r_{fl}^2 l_{fl} = K_1 K_2 v_{fl}^{id} \,, \tag{3.29}$$

then the real dimensions of a real fibril may be presented as follows:

$$r_{fl}^r = (K_1 K_2)^{1/3} r_{fl} \,,$$
$$l_{fl}^r = (K_1 K_2)^{1/3} l_{fl} \,. \tag{3.30}$$

Substituting the above equations into the formula (3.24), we find the following equation for the real heat capacity:

$$C_V^r(T) = \Delta C_V + (K_1 K_2)^{-1/3} \Phi(T) \,. \tag{3.31}$$

So, we see that to find the real character of the temperature dependence of the porous structure heat capacity one should specify the coefficients K_1 and K_2. If they do not depend on temperature, the qualitative behavior of the heat capacity will be the same as in the ideal case. In Fig. 3.3 two real situations for cases $K_1 K_2 > 1$ and $K_1 K_2 < 1$ are shown by dotted lines.

3.4 Temperature-Behavior Estimation of Porous Media Heat Capacity

Before we start explaining and describing the temperature behavior of the heat capacity of complex fiber–fibril structures, we should say some words about the heat capacity and clarify the main physical regularities of its behavior. As an example, we shall take an ordinary crystal dielectric, homogeneous in its composition and having the simplest cubic syngony (symmetry group), and agree (just here) to deal only with the isochoric heat capacity C_V.

Since the heat capacity is an equilibrium material characteristic and by its physical essence characterizes the change of internal energy (so-called heat content) for an *equilibrium* one-degree change of temperature, it is a function of two equilibrium parameters: pressure P and temperature T. We have underlined that the temperature change must be in equilibrium, i.e. comparatively slow. It means (viz. slowness of the temperature change process) that all the internal relaxation processes (leading micro-subsystems into equilibrium state and disturbed from equilibrium as result of temperature change) must have time to finish during the one-degree temperature change.

If this condition is not satisfied, the heat capacity cannot be introduced. In order to find out the dependence of C_V on temperature let us imagine a cubic dielectric with edge side equal to L, in which the number of atoms is N. If a is the interatomic distance, then $N = (L/a)^3$. As result of *equilibrium* thermal fluctuations each atom in the structure performs oscillating fluctuations near the equilibrium point x_{0i}, where the index $i = 1, 2, \ldots, N$. Let us denote these deviations by Δx_i. Then the energy of the crystal oscillating atoms with masses M will be $E = \Sigma 1/2M(\Delta \dot{x}_i^2 + \omega_i^2 \Delta x_i^2)$. If we take all the oscillations on the average to be equal and use the virial theorem, according to which the average value (with respect to time) of the kinetic energy is equal to the average value of the potential energy, i.e. $1/2M\Delta \dot{x}_i^2 = 1/2M\omega_i^2 \Delta x_i^2$, we get $E = NM\omega^2 \Delta x^2$, where ω is the average frequency of the oscillators.

Both the frequency and the shift Δx are independent statistical parameters for which the Gaussian normal law may be chosen as the distribution.

Such a simple oscillator model allows us to introduce the quantum analogue of oscillations. The corresponding formalism is the substitution of the classical Hamiltonian for the linear oscillator, $H = p_x^2/2M + 1/2M\omega^2\Delta x^2$, where the momentum of an atom is $p_x = Mv_x = Md(\Delta x)/dt$, by the Hamiltonian operator

$$H = \frac{p_x^2}{2M} + \frac{1}{2}M\omega^2 x^2 , \tag{3.32}$$

where the momentum operator $p_x = -i\hbar \partial/\partial x$ and the term Δx has been substituted by x for convenience.

Let us introduce the following two secondary-quantization operators:

$$a = \left(\frac{\hbar}{2M\omega}\right)^{1/2} \left(\frac{\partial}{\partial x} + \frac{M\omega x}{\hbar}\right) \tag{3.33}$$

and its conjugate

$$a^+ = \left(\frac{\hbar}{2M\omega}\right)^{1/2} \left(-\frac{\partial}{\partial x} + \frac{M\omega x}{\hbar}\right) . \tag{3.34}$$

If now we compose the commutator of these operators, viz. the difference $aa^+ - a^+a$, (abbr. denoted as $[a, a^+]$), then after elementary calculations we obtain

$$aa^+ - a^+a = 1 . \tag{3.35}$$

As we know, the so-called Bose operators of creation (a^+) and annihilation (a) of quasi-particles satisfy this commutation condition.

It would be so if only one oscillator were concerned. However, in a solid body the quantity of atoms is N and hence for adequate description of the dielectric properties we should take into account the whole ensemble of atomic oscillators. By means of the Hamiltonian (3.32) it is performed quite elementarily, i.e. we should introduce the sum over all oscillators. Hence we represent the Hamiltonian as

$$H = \sum_{i=1}^{N} \left[-\left(\frac{\hbar^2}{2M}\right) \left(\frac{\partial^2}{\partial x_i^2}\right) + \frac{1}{2} M\omega^2 x_i^2 \right] . \tag{3.36}$$

By means of the representation of secondary quantization (3.33) and (3.34) the equation for H in the momentum space, which is very convenient for further discussion, will be the following:

$$H = \sum_{k} \hbar\omega_k \left(a_k^+ a_k + \frac{1}{2} \right) , \tag{3.37}$$

where the law of phonon dispersion is $\omega_k = c_s k$, where c_s is the average velocity of a sound quantum, which may be introduced, for example, according to the formula $3/c_s^3 = 2/c_t^3 + 1/c_l^3$, where c_l and c_t are longitudinal and transverse sound velocities respectively, and k is the wave vector.

The operators $a_k^+ (a_k)$ characterize not abstract conjugate operators, but purely physical creation (annihilation) operators of a phonon with wave vector $\boldsymbol{k}$. For them, as well as for the operators (3.33) and (3.34) there is a more general commutation rule:

$$[a_k, a_{k'}^+] = \Delta_{kk'} , \tag{3.38}$$

where the symbol $\Delta_{kk'}$ is defined by the conditions $\Delta_{kk'} = 1$, if $k = k'$, and $\Delta_{kk'} = 0$, if $k \neq k'$.

Note as a reference that the described model is called Debye's model and due to the Hamiltonian (3.37) by means of this model the heat capacity of a crystal dielectric may be easily estimated. To demonstrate it we shall write the equation for free energy in the form:

$$F = -T \ln \sum_{k} \left(\exp \left\{ -\frac{\sum\limits_{n=0}^{\infty} \hbar\omega_k n}{T} \right\} \right) . \tag{3.39}$$

The sum over "n" is, as it is easy to notice, a simple summation series of infinitely descending geometric progression with the common ratio $q = \exp(-\hbar\omega_k/T)$.

Indeed,

$$\sum_{n=0}^{\infty} e^{-nx} = 1 + e^{-x} + e^{-2x} + e^{-3x} + \ldots = (1 - e^{-x})^{-1} .$$

Thus, after summation over "n", we get

$$F = T \sum \ln \left[1 - \exp \left(-\frac{\hbar\omega_k}{T} \right) \right] .$$

In the above formula, in order to perform the summation over the wave vectors "k", the number of phonon subsystem states should be introduced, which will be denoted as G, where

$$G = \frac{gV d^3 k}{(2\pi)^3},$$

where the multiplier $g = 2$; it accounts for the two directions of phonon polarization: longitudinal and transverse.

In such a case the formula for the transition from summation to integration over "k" is realized by means of the rule:

$$\sum(\dots) = 2V \int(\dots)\frac{d^3 k}{(2\pi)^3}.$$

Therefore, we have

$$F = 2\left[\frac{TV}{(2\pi)^3}\right] \int d^3 k \ln\left[1 - \exp\left(-\frac{\hbar\omega_k}{T}\right)\right]. \tag{3.40}$$

If now we pass over from Cartesian to spherical coordinates, substituting $d^3 k$ in the form $d^3 k = dk_x dk_y dk_z = k^2 \sin\theta d\theta d\varphi dk$ and using the identity $\int\int \sin\theta d\theta d\varphi = 4\pi$, then for the free energy we get

$$F = \left(\frac{TV}{\pi^2}\right) \int_0^{\pi/a} k^2 \ln\left[1 - \exp\left(-\frac{\hbar c_s k}{T}\right)\right], \tag{3.41}$$

where a is the interatomic distance and the integration is performed in the limits of the first Brillouin zone.

If we evaluate the integral in (3.41) by parts, we obtain

$$F = \left(\frac{\hbar c_s V}{3\pi^2}\right) \int_0^{\pi/a} \frac{k^3 dk}{[\exp(\hbar c_s k/T) - 1]}. \tag{3.42}$$

Making, at last, the substitution $\hbar c_s k/T = x$, we ultimately obtain

$$F = F(T, V) = \left(\frac{\hbar c_s V}{3\pi^2}\right)\left(\frac{T}{\hbar c_s}\right)^4 J(T), \tag{3.43}$$

where the abbreviated denotation of the integral is introduced as

$$J(T) = \int_0^{\theta_D/T} \frac{x^3 dx}{(e^x - 1)}. \tag{3.44}$$

The new parameter figuring in the upper integration limit and determined by the formula $\theta_D = \hbar c_s \pi/a$ is called the Debye temperature.

At low temperatures when the condition $T \ll \theta_D$ is satisfied, the upper limit of integration (because of the fast convergence of the integral) may be substituted by infinity, and therefore the function $J(T)$ becomes a constant, that is,

$$J(T) = \int\limits_{0}^{\infty} \frac{x^3 \mathrm{d}x}{(e^x - 1)} = \frac{\pi^4}{15}.$$

Thus, in the range of low temperatures the equation for the Helmholtz free energy is

$$F = \left(\frac{\pi^2}{45}\right) V \hbar c_s \left(\frac{T}{\hbar c_s}\right)^4. \tag{3.45}$$

So, for a solid-phase crystal structure thermodynamic the potential F has been calculated.

Further, according to the definition, the entropy is calculated as the temperature derivative from the free energy taken with the "minus" sign and at constant volume, that is, $S = -(\partial F/\partial T)_V$. Using (3.45), we find

$$S = \left(\frac{4\pi^2}{45}\right) V \left(\frac{T}{\hbar c_s}\right)^3, \tag{3.46}$$

and, therefore, the isochoric heat capacity, determined as $C_V = T(\partial S/\partial T)_V$, turns out to be $C_V = (4\pi^2/15)V(T/\hbar c_s)^3$. Introducing here the Debye temperature, we get a more convenient formula:

$$C_V = N \left(\frac{4\pi^5}{15}\right) \left(\frac{T}{\theta_D}\right)^3, \tag{3.47}$$

where the number of atoms in the crystal lattice is $N = V/a^3$.

If now in (3.47) we pass to the heat capacity of a volume unit, i.e. we take $c_V = C_V/V$, we shall get

$$c_V = \left(\frac{4\pi^5}{15a^3}\right) \left(\frac{T}{\theta_D}\right)^3. \tag{3.48}$$

For the estimation of the multiplier standing before $(T/\theta_D)^3$ we shall choose the standard interatomic distance $a = 3 \times 10^{-8}$ cm; then $(4\pi^5/15a^3) \approx 3 \times 10^{24}(1/\mathrm{cm}^3)$.

So, everything is clear with the crystal solid phase and we have assured ourselves that the heat capacity at low temperatures behaves in accordance with the traditional cubic law of Debye (3.47).

Besides the heat capacity of the solid phase we shall need the dependence of the heat capacity from the temperature in the gaseous phase, and hence we should remember that at least for a two-atomic gas (to say nothing of multi-atomic molecules) the heat capacity as an additive function is composed of three parts: (a) translational; (b) rotational; (c) oscillatory. Regarding the rotational and oscillatory parts, in the range of, say, room temperatures, (which is the most interesting for us), their contributions appear to be exponentially small and they may be neglected. But the translational part of the heat capacity connected quasi-classically with the gas molecule motion is very important for us.

For its estimation it is very convenient to use the internal gas energy E rather than the free energy (see the formula (2.63)). The connection of the heat capacity with E is:

$$C_V = \left(\frac{\partial E}{\partial T}\right)_V .$$
(3.49)

In order to find C_V, therefore, we should calculate the internal energy E.

As the kinetic energy of the translational motion of the molecules is $\varepsilon = p^2/2M$, where, as in the case of an atom in the crystal body, the molecule mass will be denoted as M, the average gas energy may be written in the form of the equation:

$$E = \frac{N_g \int \varepsilon f_p \mathrm{d}^3 p}{\int f_p \mathrm{d}^3 p} ,$$
(3.50)

where N_g is the number of gas molecules and f_p is the equilibrium Maxwell distribution function having the form

$$f_p = C \exp\left(-\frac{p^2}{2MT}\right) ,$$
(3.51)

where C is a normalization constant which may be found from the condition

$$C = \left[\int \exp\left(-\frac{p^2}{2MT}\right) \mathrm{d}^3 p\right]^{-1} .$$
(3.52)

Thus, the energy is given by

$$E = N_g \frac{\int \varepsilon \exp\{-p^2/2MT\}\mathrm{d}^3 p}{\int \exp\{-p^2/2MT\}\mathrm{d}^3 p} = N_g \frac{J_2}{J_1} .$$
(3.53)

Let us calculate the included integrals J_1 and J_2.

We shall start from J_1. If we move into a spherical system of coordinates, we shall easily obtain the following equation:

$$J_1 = \int \exp\left(-\frac{p^2}{2MT}\right) \mathrm{d}^3 p = \int_0^\infty \exp\left(-\frac{p^2}{2MT}\right) p^2 \mathrm{d}p \int_0^\pi \sin\theta \mathrm{d}\theta \int_0^{2\pi} \mathrm{d}\varphi$$

$$= 4\pi \int_0^\infty \exp\left(-\frac{p^2}{2MT}\right) p^2 \mathrm{d}p .$$

Denoting the exponent power as x, i.e. $x = p^2/2MT$, and hence expressing p, according to the relation $p = (2MT)^{1/2}x^{1/2}$, we obtain

$$J_1 = 2\pi(2MT)^{3/2} \int_0^\infty x^{3/2} e^{-x} \mathrm{d}x = 2\pi(2MT)^{3/2}\Gamma(5/2) ,$$
(3.54)

where $\Gamma(\xi)$ is the gamma function determined by the formula:

$$\Gamma(\xi) = \int_0^\infty x^{\xi-1}e^{-x}\mathrm{d}x. \tag{3.55}$$

Quite analogously for the integral J_2 we get

$$J_2 = 2\pi(2MT)^{5/2}\int_0^\infty x^{5/2}e^{-x}\mathrm{d}x = \left(\frac{\pi}{M}\right)(2MT)^{5/2}\Gamma(7/2). \tag{3.56}$$

Substituting now the derived integrals into (3.53), we may therefore determine the desired average energy of the translational motion of the gas molecules

$$E = N_g\frac{(\pi/M)(2MT)^{5/2}\Gamma(7/2)}{2\pi(2MT)^{3/2}\Gamma(5/2)} = TN_g\frac{\Gamma(7/2)}{\Gamma(5/2)} = 2.5TN_g. \tag{3.57}$$

Thus the gas heat capacity will be

$$C_V = C_{2V} = 2.5N_g. \tag{3.58}$$

If we introduce the heat capacity of the gaseous-phase volume unit according to the definition $c_v = C_{2V}/V_2$, from (3.58) we obtain

$$C_{2v} = \frac{2.5N_g}{V_2} = \frac{2.5MN_g}{MV_2} = 2.5\frac{\rho_2}{M}, \tag{3.59}$$

where ρ_2 is the gas density.

Taking (only for estimating the heat capacity c_{2v}) the density as $\rho_2 = 10^{-2}\,\mathrm{g/cm^3}$ and the molecular mass $M = 10^{-24}\,\mathrm{g}$, from (3.59) it is easy to find that $c_{2v} \approx 10^{22}(1/\mathrm{cm^3})$.

Therefore, the comparison of Eqs. (3.59) and (3.47) shows that for a porosity m of order of the difference $(1-m)$ in the case of heavy molecules of the gaseous phase and for not too low temperatures the condition $c_{1v} \gg c_{2v}$ will be always valid.

Considering in the first approximation the solid fiber–fibril structure as a crystal structure, the general expression for the total heat capacity of a volume unit according to (2.43) may be written in the form:

$$c_v = (1-m)c_{1v} + mc_{2v} \approx mc_{2v}. \tag{3.60}$$

If we now use a model of close packing of some cubic volume $V = L^3$ (L is the edge length) with solid balls of radius R, then we should take $m = 1 - \pi/6$ (for the details of obtaining this relation the reader is referred to the text before the formula (2.58)). Then, instead of (3.60), heat capacity is expressed as

$$c_v \approx \left(\frac{\pi}{6}\right)c_{1v} = \left(\frac{2\pi^5}{45a^3}\right)\left(\frac{T}{\theta_D^*}\right)^3, \tag{3.61}$$

where the new parameter θ_D^* introduced here is some effective temperature (the prototype of the Debye temperature in a crystal dielectric) for the fibril structure.

As we can see in Fig. 3.5, the experimental data show that there is maximum heat capacity at a temperature of about $100°C$ (373 K) for practically all the paper and cellulose samples. Thus, our task is to give an adequate interpretation of the experimental results based on the already available partial information on the heat capacity of a unit volume of ideal porous medium (formula (3.61)). Here attention should be paid to the fact that the experiment actually concerns the heat capacity of a unit mass of a sample. In order to obtain the relation between the heat capacity of a unit volume and the heat capacity of a unit mass let us introduce the heat capacity of a unit mass as the relation $C_M = C_V/M_0$, where M_0 is the entire mass of the sample. Then

$$C_M = \frac{C_V}{M_0} = \left(\frac{C_V}{V}\right)\left(\frac{V}{M_0}\right) = \frac{c_v}{\rho} = \frac{(1-m)c_{1v} + mc_{2v}}{(1-m)\rho_1 + m\rho_2}, \tag{3.62}$$

where ρ_1 and ρ_2 are the solid and gaseous densities, respectively.

In the experiment the sample was thermally isolated from external heat sources and therefore it is held that we deal with an adiabatic process. In an adiabatic process the relation between the temperature and the volume is expressed by the known equation $TV^{\gamma-1} = C$, where C is a constant and the adiabatic exponent is $\gamma = C_P/C_V$, where C_P is the isobaric heat capacity. Solving this equation with respect to V we find

$$V = \frac{C}{T^{1/\gamma-1}} = \frac{C}{T^\alpha}, \tag{3.63}$$

where the index of power $\alpha = 1/\gamma - 1 = C_V/N$.

As we are interested in the relation between the volume and the temperature only in the gaseous phase, then, assigning the lower index "2" from (2.63) to the corresponding parameters we obtain

$$V_2 = \frac{C}{T^\alpha}, \tag{3.64}$$

where $\alpha = C_{2V}/N_2 = c_{2v}^*$. Here the new heat capacity c_{2v}^* is introduced, which is the heat capacity of one molecule.

If now we introduce the gas density using the formula $\rho_2 = M_2/V_2$, where M_2 is its mass, then from (3.64) we obtain the relation between density and temperature:

$$\rho_2 = c^* T^{c_{2v}}, \tag{3.65}$$

where C^* is some re-normalized constant.

As a result, by means of the relation (3.65) the heat capacity of a unit mass may be written as

$$C_M = \frac{(1-m)c_{1v} + mc_{2v}}{(1-m)\rho_1 + m\rho_2} \approx \frac{c_{1v}}{1 + [m/(1-m)]C^*(T^{c_{2v}}/\rho_1)}, \tag{3.66}$$

where we have used the condition $c_{1v} \gg c_{2v}$.

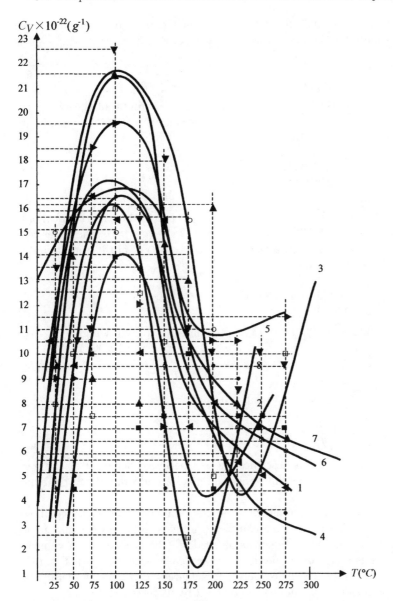

Fig. 3.5. Experimentally observed temperature dependencies of the cellulose heat capacity. The solid lines are fits to the experimental points. Curves 1–8 correspond to Table 3.2 and characterize the particular substance (1: ◀, 2: ■, 3: ▼, 4: •, 5: ▶, 6: ○, 7: ▲, 8: □)

Substituting here instead of c_{1M} its explicit expression from (3.61), we find the formula for the desired heat capacity of a unit mass:

$$C_M = \left(\frac{2\pi^5}{45a^3}\right)\left(\frac{T}{\theta_D^*}\right)^3 \frac{1}{1 + [m/(1-m)]C^*(T^{c_{2v}}/\rho_1)} . \tag{3.67}$$

Setting now the derivative dC_M/dT equal to zero, we find that the extremum is possible only in the case when c_{2v} is larger than three, and the maximum point lies at the temperature

$$T_{\max} = T^* \left[\frac{3(1-m)}{(c_{2v}-3)m}\right]^\nu , \tag{3.68}$$

where $\nu = 1/c_{2v}$.

Let us choose T^* from the condition of reference to the experiment. For it we shall take $T_{\max} \approx 100°\text{C} = 373$ K, $c_{2v} = 7/2$, $1 - m = \pi/6$ (the latter condition defines the porosity of a closely packed system of fibrils approximated by spheres) and as a result we get

$$T^* = 373 \left[\frac{(6-\pi)}{6\pi}\right]^{2/7} = 217.6\,\text{K} .$$

Thus, the dependence (3.67) at $T_{\max}$ determined in (3.68) will give

$$C_M = \left(\frac{2\pi^5}{45a^3}\right)\left(\frac{T}{\theta_D^*}\right)^3 \frac{1}{1 + [m/(1-m)](T/T^*)^{c_{2v}}} . \tag{3.69}$$

Taking $m/(1-m) = 6/\pi - 1 = 0.91$, we find

$$C_M = \left(\frac{2\pi^5}{45a^3}\right)\left(\frac{T}{\theta_D^*}\right)^3 \frac{1}{1 + 0.91(T/T^*)^{c_{2v}}} . \tag{3.70}$$

The obtained equation describes the temperature behavior of the heat capacity of a porous sample unit mass and gives an adequate fit to the experiments illustrated in Fig. 3.5.

Thus, the conclusion issuing from the functional dependence (3.69) shows the domination in pores of rather complex (and most probably organic) molecules with a large number of degrees of freedom (the isochoric heat capacity per one gaseous-phase molecule must, in any case, exceed three). From the dependence (3.69) it is therefore seen that the larger is c_{2v}, the more abruptly the right-hand side of the dependence $C_M(T)$ will decrease.

It should be noted as well, that the formulae (3.60) and (3.61) obtained a little earlier qualitatively describe the analytical dependence of the specific heat capacity of a unit volume of a porous dielectric on the temperature, to which Figs. 3.4 and 3.5 may, with some allowance, be compared. The formula (3.70) for the heat capacity of a unit mass confirming the results of experiment (illustrated, in turn, by Fig. 3.5), besides other things, gives the correct physical interpretation of the obtained results.

Here it is worthwhile to pay attention to such a notable point. The thing
is that the behavior of the heat capacity in the form of a linear function
of T, which is illustrated by Fig. 3.5, may be explained by the cubic dependence (3.61) with rather considerable tolerance. Following the logic, as we are
dealing with extremely complex structures such as paper and cellulose, we
cannot in one word answer the question, why in the quoted experiment the
tendency to linear growth of c_v with T was observed. And the problem is not
how to "adjust" the physical essence of the phenomenon to the interpretation
of the given experiment, but at least partially to gain understanding by re-
presenting such complex substances in terms of the maximum simplification
of their physical model. The first thing which suggests itself as a prompt in
interpreting the linear relation of c_v with T, is the so-called two-level systems
for which (in accordance with the Anderson–Philips model, which is valid for
a glass-like substance) the heat capacity will really grow linearly with temperature, but only in the range of very low temperatures. In the range of high
temperatures this is not the case. So far the given problem remains open.

3.5 Comparison with Experiment

In order to clear up the agreement of the theoretical description of the heat
capacity $C_V(T)$, whose behavior is illustrated in Fig. 3.5, with reality, a simple
but labor-consuming experiment was conducted with the purpose of studying the behavior of $C_V^P(T)$ for different types of cellulose and paper in the
temperature range from 50°C up to 275°C (see Figs. 3.4 and 3.5).

Not to fatigue the reader by philosophical speculations we present Table 3.2 with the measured values of the heat capacity.

3.6 Mechanical Strength of the Porous Structure: Calculation of the Break Force and Its Dependence on the Substance Density

If the model representation of the porous dielectric structure is available, the
task to calculate the break force does not seem to be too complicated. Its
significance refers to the preliminary analysis of the phenomenological physics
and only after it the search for a method of theoretical description.

Let some force F_p be applied to the dielectric. As the result of the local
action of this force in the place of its application the characteristic break
contour will appear (see Fig. 3.6b). The question arises: what is the intrinsic
cause of the counteraction to the external force? Within the framework of the
model submitted in Sect. 2.2 it is quite clear that the force F_p must overcome
the internal interaction of the f-points. Indeed, the fibril interaction energy
(see Fig. 2.5), as it has already been stated, has an electromagnetic nature

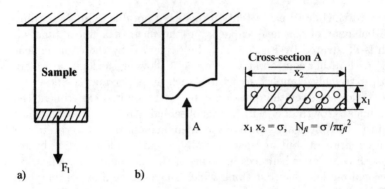

Fig. 3.6. Schematic representation of the experiment of mechanical strength study (**a**) and the structure after disrupture (**b**). Section A shows the chaotic arrangement of the fibrils.

and is determined by Van der Waals forces on condition that the distance between the interacting bodies is of order of 10^{-4} cm or less (for more details of the nature of these forces refer to the monograph [14]). Therefore, for fibrils whose characteristic diameters are about 10^{-4}–10^{-5} cm it indirectly shows the correct scale of the action of these forces. According to Sect. 3.2. the force of "cohesion" of fibrils is

$$F_{\text{int}}^{(1)} = \left(\frac{\pi^2 Z_{\max}}{240}\right)\left(\frac{\hbar c}{S^4}\right)\left[\frac{(\varepsilon - 1)}{(\varepsilon + 2)}\right] N\sigma J(\varepsilon), \tag{3.71}$$

where S is the distance between the fibril axes (see Fig. 3.7), σ is the break section, and N is the complete number of f-points in the break section. The function $J(\varepsilon)$ very weakly affects the interaction force and over the whole real range of values ε is approximately equal to 0.4.

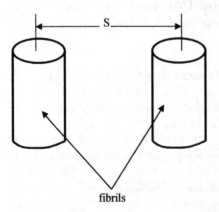

Fig. 3.7. Image of fibrils arranged at distance S from each other and interacting in accordance with the Van der Waals law

In order to use the shown dependence for the estimation of the break force the number of f-points in the section σ should be calculated. From Fig. 2.5 and Fig. 3.6b we get

$$N = \frac{\sigma}{\pi R^2} = \frac{\sigma}{8\pi r_{fl}^2}$$

and hence

$$F_b^* = F_{int} = 6 \times 10^{-4} \left(\frac{Z_{max} \hbar c \sigma^2}{S^4 r_{fl}^2} \right) \left[\frac{(\varepsilon - 1)}{(\varepsilon + 2)} \right]^2 .$$

Keeping the distance S approximately equal to $6r_{fl}$ and assuming that $v_{fl} = \pi r_{fl}^2 l_{fl}$, we find

$$F_b^* = 1.4 \times 10^5 \left[\frac{Z_{max} \rho^3 c \sigma^2 l_{fl}}{(1 - \mu)^3 S^4 M_{fl}^3} \right] \left[\frac{(\varepsilon - 1)}{(\varepsilon + 2)} \right]^2 . \tag{3.72}$$

Thus we see that the break force F_b^* is proportional to the cubic density. Let us estimate F_b^*. Taking $Z_{max} = 9$, $\rho/(l - \mu) = \rho_{fl} = 1.55\,\mathrm{g/cm^3}$, $\hbar = 10^{-27}$ erg s, $c = 3 \times 10^{10}$ cm/s, $l_{fl} = 10^{-3}$ cm, $M_{fl} = 5 \times 10^{-13}$ g, and $\varepsilon = 1.2$, we have $F_b^* = 0.9\,\mathrm{kg}$.

It should be noted that if we do not limit ourselves to the ordered fibril structure and assume a chaotic fiber orientation, the result (3.72) will change and the qualitative relation between the break force and the density will be linear rather than cubic, which may be seen, in particular, from some experiments (e.g. see [15–18] and the monograph [19]). Let us try to describe these cases.

Let the fibrils be oriented randomly in the structure volume and make an angle θ_i with the specified direction space. Then, according to Parsegian [20] the interaction force between two arbitrarily chosen nearest neighbors may be presented in the following way:

$$F_2(l, \theta_i) = \frac{2N_1 N_2 C(\theta_i)}{9l^2} , \tag{3.73}$$

where $N_{1,2}$ is the average number of fibrils per unit of the cross-sectional area of the system under consideration and l is the shortest distance between the fibrils.

The function

$$C(\theta_i) = 9 \left(\frac{S_1 S_2}{512\pi} \right) \sum_{n=0}^{\infty} \left\{ \frac{(18 + \cos 2\theta_i)(\varepsilon_{1a} - \varepsilon_3)(\varepsilon_{2a} - \varepsilon_3)}{(\varepsilon_{1a} - \varepsilon_3)(\varepsilon_{2a} - \varepsilon_3)} \right.$$

$$+ 0.5(6 - \cos \theta_i) \left[\frac{(\varepsilon_{1t} - \varepsilon_3)(\varepsilon_{1a} - \varepsilon_3)}{\varepsilon_3(\varepsilon_{1a} + \varepsilon_3)} + \frac{(\varepsilon_{1t} - \varepsilon_3)(\varepsilon_{2a} - \varepsilon_3)}{\varepsilon_3(\varepsilon_{2a} + \varepsilon_3)} \right]$$

$$\left. + 0.25(2 + \cos 2\theta_i) \frac{(\varepsilon_{1a} + \varepsilon_3)(\varepsilon_{2a} + \varepsilon_3)}{\varepsilon_3^2} \right\} , \tag{3.74}$$

where the areas of the fibril cross-sections $S_{1,2} = \pi r^2_{1,2fl}$, $\varepsilon_{1a,t(2a,t)}$ are the main values of the dielectric permeability tensor in the perpendicular (index "a") and parallel (index "t") directions relative to the specified axis Z for the first and the second fibrils, respectively, and ε_3 is the dielectric permeability of the medium filling the space between the fibrils. For air we take $\varepsilon_3 = 1$.

We shall underline once more that unlike the interaction force (3.71) Eq. (3.73) takes into account only the pair-wise interactions.

Now, averaging the function $C(\theta_i)$ over angles, which is confined to its simple integration with respect to all θ in the range from 0 to π with the weight factor "one" we find

$$\langle C \rangle = C = 9 \left(\frac{S_1 S_2}{512\pi} \right) T \sum_{n=0}^{\infty} \left\{ \frac{18(\varepsilon_{1a}(\omega) - 1)(\varepsilon_{2a}(\omega) - 1)}{(\varepsilon_{1a}(\omega) + 1)(\varepsilon_{2a}(\omega) + 1)} \right.$$

$$+ 3 \left[\frac{(\varepsilon_{1t}(\omega) - 1)(\varepsilon_{1a}(\omega) - 1)}{\varepsilon_{1a}(\omega) + 1} + \frac{(\varepsilon_{1t}(\omega) - 1)(\varepsilon_{2a}(\omega) - 1)}{\varepsilon_{2a}(\omega) + 1} \right]$$

$$\left. + 0,5(\varepsilon_{1a}(\omega) - 1)(\varepsilon_{2a}(\omega) - 1) \right\}, \tag{3.75}$$

where the frequency $\omega = 2\pi nT/\hbar$. Thus the break force will be

$$F_{2b} = \frac{S_1 S_2 N_1 N_2 l_{fl} C^*}{256\pi l^2}, \tag{3.76}$$

where renormalized the constant $C^* = 2C/9S_1 S_2$.

Now we shall reduce Eq. (3.76) to a more practical form. Indeed, taking the masses of all the fibrils to be the same and equal to M_F we find the relation for the break force of a porous medium to be

$$F_{2b} = q_1 \rho, \tag{3.77}$$

where the coefficient

$$q_1 = \frac{S_1 S_2 N_1 N_2 l_{fl}^2 C^*}{256\pi M_{fl}}. \tag{3.78}$$

Let us estimate the value F_{2b}. Taking $T = 10^{-14}$ erg, $\rho = 1.3\,\mathrm{g/cm^3}$, $l_{fl} = 10^{-3}$ cm, $r_{fl} = 10^{-4}$ cm, $M_{fl} = \rho_{fl} v_{fl} = \rho_{fl} \pi r_{fl}^2 l_{fl} = 5 \times 10^{-14}$ g, $S_1 N_1 = S_2 N_2 = 10^4$, we find $F_{2b} = 5\,\mathrm{kg}$. The given estimation is in good agreement with the experiments [15–18].

It is obvious that the whole break force should be composed of two components F_b^* and F_{2b}. Indeed, according to (3.72) and (3.77) we get

$$F_b = F_b^* + F_{2b} = q_1 \rho + q_2 \rho^3, \tag{3.79}$$

where q_2 is determined by (3.72).

In Fig. 3.8 the dependence of the break force on the structure density is shown. The points correspond to the experimental data.

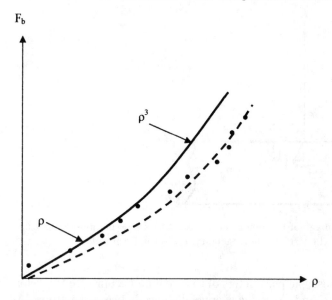

Fig. 3.8. Dependence of the break force on the porous structure density. The solid line is theory, the dotted line is experiment.

3.7 Calculation of the Components of the Tension Tensor Near Macroheterogeneity

In the practical production of a denser porous medium, which is achieved by exposure of the structure surface to the compression force (rolling), there always exists a probability of macroheterogeneities in some local place on the surface of the sample due to the inhomogeneity of the press surface. This local place is the center of the tension whereby the structure break would most probably take place under mechanical action (see the previous section).

In such a local place the energy of the internal stress is higher than elsewhere and consequently at the action of the break force there is a kind of additional force acting along the direction of the external load. That is,

$$F_{\text{load}} = F_{\text{break}} - F_{\text{loc}} .$$

Let us now estimate the elastic deformation energy concentrated near this local heterogeneity.

Imagine, according to Fig. 3.9, that there is some recess (macrodefect) with radius R on the dielectric surface with thickness δ. The circles R_0 are f-points, each consisting of Z_{max} fibrils.

In order to calculate the energy of this tension, first of all it is necessary to choose reasonably the deformation tensor of such a porous structure and then to calculate its dependence on the two-dimensional radius r.

The continuum approximation in the theory of elasticity [21] justifies the validity of the inequality $\delta x \gg \langle a \rangle$, where δx is the domain of characteristic

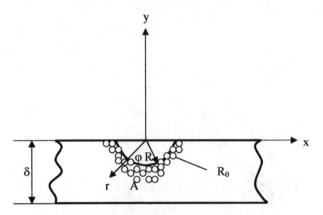

Fig. 3.9. Symbolic image of the deformation resulting from the non-ideal surface of a rolling mill. R is the size of inhomogeneity, r_{fl} is the fibril radius.

changes of the macroscopic parameters of the medium (in our case the deformation tensor $u_{ik}(r)$) and $\langle a \rangle$ is the average distance between the atoms. As it is impossible to deal with interatomic distances in the case under description, the question arises: what will play the role of the parameter $\langle a \rangle$ in the framework of our problem?

As the characteristic range of the deformation potential change has dimensions of the order of the fibril diameter, then in terms of the f-points model, it is natural to suppose that the value $2r_F$ will act as $\langle a \rangle$. We shall make this supposition. So, we consider that the following condition is valid:

$$\delta x \gg 2r_{fl} \, . \tag{3.80}$$

Now we have the grounds to speak about the continuum approximation and can use the general equations of theory of elasticity [21, 22].

For the medium shift vector $u(r)$, taking account of the external local pressure (P_k) on the specified area of the surface, the following equation may be written:

$$\Delta u + (1 - 2\sigma)^{-1} \mathrm{grad\,div} u = -\frac{4P_k R_2(1 + \sigma)n}{2LEr^2} \, , \tag{3.81}$$

where the unit vector of the normal is $n = \{n_r, n_\varphi\} = \{0, \sin\varphi\}$, r is a running coordinate, E is the Young's modulus of the dielectric, σ is Poisson's coefficient, and L is the defect extent perpendicular to the surface in Fig. 3.9. The deformation tensor is related to the shift vector by the known equation:

$$u_{ik} = 0.5 \left(\frac{\partial u_i}{\partial x_k} + \frac{\partial u_k}{\partial x_i} \right) \, . \tag{3.82}$$

In order to solve Eq. (3.81) it is convenient to pass over to the cylindrical coordinate system. In the two-dimensional case this equation may be written down in projections as two linearly independent equations. That is,

$$
\left\{
\begin{aligned}
& \left(\frac{1}{r}\right)\left(\frac{\partial}{\partial r}\right)\left(\frac{r\partial u_r}{\partial r}\right) + \left(\frac{1}{r^2}\right)\left(\frac{\partial^2 u_r}{\partial \varphi^2}\right) \\
& + (1 - 2\sigma)^{-1}\left(\frac{\partial}{\partial r}\right)\left(\frac{1}{r}\right)\left[\left(\frac{\partial(ru_r)}{\partial r}\right) + \frac{\partial u_\varphi}{\partial \varphi}\right] = 0 \\
& \left(\frac{1}{r}\right)\left(\frac{\partial}{\partial r}\right)\left(\frac{r\partial u_\varphi}{\partial r}\right) + \left(\frac{1}{r^2}\right)\left(\frac{\partial^2 u_\varphi}{\partial \varphi^2}\right) \\
& + (1 - 2\sigma)^{-1}\left(\frac{1}{r^2}\right)\left(\frac{\partial}{\partial \varphi}\right)\left[\left(\frac{\partial(ru_r)}{\partial r}\right) + \frac{\partial u_\varphi}{\partial \varphi}\right] = -4P_k R^2 (1 + \sigma) \\
& \hspace{9cm} \times \frac{\sin \varphi}{LEr^2} \,.
\end{aligned}
\right.
$$

$$(3.83)$$

Let us make a simple assumption concerning the components u_r and u_φ. We shall assume the component u_r to be a function only of r and the component u_φ a function only of the angle φ.

In this case the set of Eqs. (3.83) will become much simpler and we easily find

$$
\left\{
\begin{aligned}
& \left(\frac{1}{r}\right)\left(\frac{\partial}{\partial r}\right)\left(\frac{r\partial u_r}{\partial r}\right) + (1 - 2\sigma)^{-1}\left(\frac{\partial}{\partial r}\right)\left(\frac{1}{r}\right)\left[\frac{\partial(ru_r)}{\partial r}\right] = 0 \,, \\
& \frac{\partial^2 u_\varphi}{\partial \varphi^2} = -2P_k R^2 (1 + \sigma)(1 - 2\sigma)\frac{\sin \varphi}{LE} \,.
\end{aligned}
\right.
$$

$$(3.84)$$

From the second equation immediately follows the solution

$$
u_\varphi = 2P_k R^2 (1 + \sigma)(1 - 2\sigma)\frac{\sin \varphi}{LE} \,,
\tag{3.85}
$$

and the first equation is reduced to the form

$$
2(1 - \sigma)r^2 u_r'' + r u_r' - u_r = 0 \,.
\tag{3.86}
$$

Its solution with a real physical meaning is

$$
u_r = \frac{A}{r^{1/2(1-\sigma)}} \,.
\tag{3.87}
$$

The constant A is found from the condition $u_r = r_{fl}/2$ at $r = R$. Therefore,

$$
u_r = \frac{0.5 r_{fl}}{(R/r)^{1/2(1-\sigma)}} \,.
\tag{3.88}
$$

Consequently, according to (3.82), the deformation tensor components different from zero are

$$
\left\{
\begin{aligned}
& u_{rr} = -\left[\frac{r_{fl}}{4(1 - \sigma)r}\right]\left(\frac{R}{r}\right)^{1/2(1-\sigma)} \,, \\
& u_{\varphi\varphi} = 2P_k R^2 (1 + \sigma)(1 - 2\sigma)\frac{\sin \varphi}{(1 - \sigma)LEr} \,.
\end{aligned}
\right.
$$

$$(3.89)$$

The equations obtained allow us to get comprehensive physical information about the A region. With this purpose we shall calculate the potential energy stored in the place of local tension. From [21] we have the equation for it:

$$\Omega = \mu \left(\frac{u_{ik} - \delta_{ik} u_{ll}}{3} \right)^2 + 0.5 K u_{ll}, \tag{3.90}$$

where Ω is free energy density related to one cubic centimeter, and μ and K are the moduli of shift and compression, respectively. Their relation to Poisson's ratio and Young's modulus is as follows:

$$\begin{cases} K = \dfrac{E}{3}(1 - 2\sigma), \\[2mm] \mu = \dfrac{E}{2}(1 + \sigma). \end{cases} \tag{3.91}$$

Substituting into (3.90) the equations for u_{rr} and $u_{\varphi\varphi}$, we find the energy density sought for:

$$\begin{aligned} \Omega &= \left(\frac{K}{2} + \frac{2\mu}{3} \right)(u_{rr}^2 + u_{\varphi\varphi}^2) \\[2mm] &= \left(\frac{K}{2} + \frac{2\mu}{3} \right)\left\{ 4P_k^2 R^4 (1+\sigma)^2 (1-2\sigma)^2 \frac{\sin^2\varphi}{(1-\sigma)^2 L^2 E^2 r^2} \right\} \\[2mm] &\quad + \left[\frac{r_{fl}^2}{16(1-\sigma)^2 r^2} \right] \left(\frac{R}{r} \right)^{1/(1-\sigma)}, \end{aligned} \tag{3.92}$$

and, finally, since the complete energy is integral over the A region is

$$\varepsilon_{\text{loc}} = \int\limits_{V_A} \Omega \mathrm{d}^3 x,$$

in the cylindrical coordinate system we have

$$\varepsilon_{\text{loc}} = L \int\limits_{R}^{\delta} \int\limits_{0}^{\pi} \Omega(r, \varphi) r \mathrm{d}r \mathrm{d}\varphi.$$

Substituting here Eq. (3.92) and making a simple integration we find

$$\varepsilon_{\text{loc}} = \pi L \left(\frac{K}{2} + \frac{2\mu}{3} \right) \tag{3.93}$$

$$\times \left\{ \frac{2[P_k^2 R^4 (1+\sigma)^2 (1-2\sigma)^2 \ln(\delta/R)]}{(1-\sigma)^2 L^2 E^2} + \frac{r_{fl}^2}{16(1-\sigma)^2} \left[1 - \left(\frac{R}{\delta} \right)^{1/(1-\sigma)} \right] \right\}.$$

Now, taking into account the energy spent on pressing out the volume A, Eq. (3.94) must include the addend $\pi R^2 L P_k$, i.e. the complete energy is

$$\varepsilon_F = P_k \pi R^2 L + \varepsilon_{\text{loc}}. \tag{3.94}$$

From the equation obtained we may find, in particular, the linear dimension of the defect L for which the complete energy is a minimum. From the condition $\partial \varepsilon_F / \partial L = 0$, using also the inequality $P_k R^2 \gg E r_{fl}^2$, we find

$$L_{min} = R \left\{ \left[2P_k \ln \left(\frac{\delta}{R} \right) \right] \frac{(1+\sigma)(1-2\sigma)}{E(1-\sigma)} \right\}^{1/2}. \tag{3.95}$$

So, we come to the first conclusion. As L_{min} is proportional to $(P_k)^{1/2}$, the higher the pressure, the more extensive must be the defect. For this value of L_{min} the minimum energy is

$$\varepsilon_{Fmin} = 2P_k \pi R^3 \left\{ \left[2P_k \ln \left(\frac{\delta}{R} \right) \right] \frac{(1+\sigma)(1-2\sigma)}{E(1-\sigma)} \right\}^{1/2}. \tag{3.96}$$

Thus, the minimum energy behaves as $P_k^{3/2}$. Let us make one more analysis. We shall calculate the minimum deformation. Putting the derivative over R from (3.94) equal to zero and solving it with respect to R, we have

$$R_{min} = \left\{ \frac{r_{fl}^2 L^2 E^2}{128 P_k^2 \delta^{1/(1-\sigma)}} (1+\sigma)^2 (1-2\sigma)^2 \ln \left(\frac{\delta}{R} \right) \right\}^{(1-\sigma)/3-4\sigma)}. \tag{3.97}$$

Therefore

$$\varepsilon_{loc\,min} = \left[\frac{\pi L r_{fl}^2 E}{32} (1+\sigma)(1-2\sigma) \right]$$

$$\times \left\{ 1 - \left(\frac{R_{min}}{\delta} \right)^{1/(1-\sigma)} \frac{(3-4\sigma)}{(1-\sigma)} \right\}. \tag{3.98}$$

From the obtained equation we can see that the local deformation energy changes as the value proportional to the difference $P_k^{1/2} - D/P_k^\alpha$, where according to (3.98) D is a function of the parameters σ, δ, E, and the index of power $\alpha = (1-\sigma)/2(3-4\sigma)$. So, we can formulate the second conclusion: the deformation inside the structure increases much less than the deformation along the surface (remember that ε_p is proportional to $P_k^{3/2}$).

Concluding this section and Chap. 3 we cannot help noting the great importance of the mechanical strength of porous dielectrics. Indeed, when it concerns, say, the electric breakdown of capacitor paper, the surface defects are of prior importance because in these local macroheterogeneities the electric breakdown occurs. In these places there is a tendency to accumulation of the electric charge. This phenomenon will be discussed in detail in the next chapter.

4 The Theory of Fast Nonstationary Phenomena in Porous Dielectrics

Two very important concepts – the electric breakdown field and the heat breakdown temperature – should be regarded as important physical characteristics of "purity" of a dielectric surface. The fact is that both these parameters allow us to find that weakest place in which destruction is most probable. Indeed, as we pointed out in Sect. 3.6, if as a result of mechanical treatment small defects (of micrometer size) appear on the structure surface, then, on the one hand, they will form the local area of elastic tension and on the other hand, on application of an electric field, they will form the area of electric charge concentration. Therefore, the presence in real (not hypothetical) samples of any heterogeneous composition can lead in practice to unpredictable consequences and untimely failure of some elements and mechanisms used, for example, in electrical engineering. That is why one of the tasks of the present chapter is to give a comprehensive analysis of the physical causes of breakdown and then, based on these results, to give a number of practical recommendations contributing to the improvement of the "resistance" of a porous sample to the electric breakdown field.

Let the electric field $\boldsymbol{E}$ be applied perpendicularly to the surface of a two-dimensional porous structure. When the field $\boldsymbol{E}$ reaches the critical value, as has been shown by numerous experiments, the electric breakdown occurs not immediately but after some time τ, which is described as the time of preparation of the internal structure to the breakdown. In process of this preparation specific bangs are heard in the dielectric volume showing the origination of the pre-breakdown situation. The breakdown itself lasts for a very short time and in order of magnitude it is $\tau_{br.} = \delta/\nu_e$, where δ is the film thickness and ν_e is the average speed of the electrons in the stream. In the breakdown area (see [23]) a burnt-out point is formed with a characteristic diameter (or, as they say, a confidence interval upon numerous experiments) from 0.05 cm up to 0.5 cm. We shall show that the process of burning out is the secondary effect. As the electric breakdown is necessarily accompanied by processes of internal "friction", as a result a great quantity of heat should be released at a rate $\partial T/\partial t = \alpha(T_e - T)$, where $\alpha = C_V^{(e)}/\tau_T$ is the coefficient of heat transfer from the breakdown area to the "cold" volume of the porous dielectric at temperature T, the heat transfer time is τ_T and the electron heat capacity is $C_V^{(e)}$. On the other hand, since the heat production per unit time is

$$\frac{dQ}{dt} = \int q \mathrm{d}\boldsymbol{S} = - \int \int \kappa_e \left(\frac{\partial T}{\partial t}\right) r \mathrm{d}r \mathrm{d}\varphi = -2\pi\kappa_e \delta(T_e - T),$$

where T_e is the electron temperature, κ_e is the electron thermal conductivity and therefore $\alpha = 2\pi\kappa_e\delta$. Hence, the heat transfer time is

$$\tau_T \cong \frac{C_V^{(e)}}{2\pi\kappa_e\delta}. \tag{4.1}$$

Comparing the obtained equation with τ_{br} on condition that $\tau_{br} \ll \tau_T$, we find the relation for the dielectric thickness is

$$\delta^2 \ll \frac{C_V^{(e)} v_e}{2\pi\kappa_e}.$$

Then, because $\kappa_e = \chi_e C_V^{(e)}/V_{\text{breakdown}}$, where $V_{\text{breakdown}} = \pi r_{br}^2 \delta$, r_{br} is the average breakdown radius (less than the radius of burning out!), χ_e is the electron thermal diffusion, equal in order of magnitude to the product $l_e v_e$, where l_e is electron free path. Therefore, for the relation $\tau_{br} \ll \tau_T$ to be valid, the following inequality should be valid:

$$\delta < \frac{r_{br}^2}{l_e}. \tag{4.2}$$

Thus, we have proved that the thermal breakdown takes place after the electric breakdown. Really, the inequality (4.2) always holds due to the fact that electron free path is very short.

By means of the above analysis we can estimate one more extremely important physical parameter. It concerns the time of electron collision τ_{coll} defined by the relation $\tau_{\text{coll}} = l_e/v_e$. For a purely technical application this time is insignificant; it is only important to understand the microscopicity of the breakdown process! So, since

$$\kappa_e = \frac{\chi_e C_V^{(e)}}{V_{br}} = \frac{C_V^{(e)}}{r_{br}^2 \delta},$$

from formula (4.1) we have

$$\tau_T = \frac{r_{br}^2}{\chi_e}. \tag{4.3}$$

substituting here the gas-kinetic expression for $\chi_e = l_e v_e = v_e^2 \tau_{\text{coll}}$, and recalling that $v_e = \delta/\tau_{br}$, we find

$$\tau_T = \frac{r_{br}^2 \tau_{br}^2}{\tau_{\text{coll}}\delta^2}. \tag{4.4}$$

So, the condition $\tau_{br} \ll \tau_T$ will also be valid when

$$\tau_{\text{coll}} < \left(\frac{r_{br}}{\delta}\right)^2 \tau_{br}. \tag{4.5}$$

Usually (see [23]) $r_{br} = 10^{-5}$ cm and the dielectric thickness $\delta = 10^{-3}$ cm; therefore the inequality (4.5) gives the upper estimate for $\tau_{coll} < 10^{-4}\tau_{br}$.

As a rule, $\tau_{br} = 10^{-5} - 10^{-8}$ s. And therefore $\tau_{coll} < 10^{-9} - 10^{-12}$ s. For electron processes such periods of time correspond to reality (see [24, 25]).

4.1 Physical Concept of Electric and Thermal Breakdown

Before electric breakdown takes place, radical changes in the sample volume occur followed by the disturbance of the local electric strength. Just after switching on the electric field a very weak direct electric current is sustained in the dielectric continuously, and as soon as the value of this current approaches the critical point, breakdown occurs. The pre-breakdown current, due to its low value, does not damage the internal structure of the substance. Because, according to Ohm's law, the current is $j = \sigma E$ and the conductivity σ is proportional to the squared concentration of the carriers n^-, at low concentrations the current will be low. At the concentration increase which is followed by increase of the electric field E, the current also increases and approaches the limiting threshold j_{cr}. In dielectrics the value of the conductivity corresponds in order of magnitude to the values varying about $10^{-12}\mathrm{s}^{-1}$, and it is due to the small value of all the parameters that the critical value of the breakdown field is high. As soon as the field E approaches the threshold value E_{cr}, the breakdown current increases abruptly and cannot be described by the differential Ohm's law. In this case the empiric Pull's formula "works":

$$j = j_0 \exp(A_1 E) \tag{4.6}$$

or the theoretical relation obtained by Frenkel

$$j = j_0 \exp(A_2 E^{1/2}), \tag{4.7}$$

where $A_{1,2}$ are constants.

From these two qualitatively similar formulae we can see that in a rather narrow range of E values the current abruptly rises.

It is clear, however, that the increase of j cannot be infinite and hence at definite values $E = E^*$ (the rigorous calculation of parametric dependence of E^* is given in Sect. 4.2) the current gradually decreases and tends to the relation

$$J = A_3 E^{1/2}, \tag{4.8}$$

where A_3 is the constant calculated rigorously in Sect. 4.2.

The ordinary breakdown field of porous dielectrics such as cellulose or capacitor paper lies in the range from 1.5×10^5 up to 7×10^5 V/cm. In solid crystal substances and in glasses the breakdown field is roughly 10^6–10^7 V/cm (see, for example, the monograph [26]). Qualitatively the decrease of the breakdown field in porous substances, unlike crystal ones, is quite clear.

Indeed, as the porous structure consists of the interlacing of free volumes and fibrils, to a rough approximation the breakdown field may be estimated from the empirical formula:

$$E_{br} = [m(1 - m)E_{br}^{(fl)} E_{br}^{(p)}]^{1/2},$$ (4.9)

where $E_{br}^{(fl)}$, $E_{br}^{(p)}$, are the breakdown fields of the fibril structure and the gas-phase, respectively, and m is the porosity.

Let us estimate the value of E_{br}. According to the experiments described in [27, 28] for a thickness $\delta = 2 \times 10^{-4}$ cm

$$E_{br}^{(p)} = 1.36 \times 10^5 \text{ V/cm}.$$

According to [29]

$$E_{br}^{(fl)} = 4 \times 10^6 \text{ V/cm}.$$

At maximum porosity $m = 0.5$ and from the formula (4.9) it follows that $E_{\text{br max}} = 3.7 \times 10^5$ V/cm.

For capacitor paper according to measurements described in [30] where $E_{br}^{(\text{paper})} = 2.5 \times 10^5$ V/cm. It is seen that $E_{br}^{(\text{paper})}$ does not exceed the value $E_{\text{br max}}$.

It means that, firstly, the empirical formula (4.9) gives, though roughly, agreement with experiments and the exact value of the porosity in the samples is not known; secondly, it allows us to consider the problem of the calculation of the electric breakdown field of porous media from a qualitatively different point of view.

Indeed, the calculation methods for E_{br} must include a number of consequent operations and mainly the determination of the gas-phase composition filling the pores on condition that m is known – the self-consistent approach for the determination of m will be described in detail in Sect. 5.7.

Now we shall say some words about the thermal breakdown. The first elementary theory of this phenomenon was developed by Wagner [31]. It stimulated further improvements and development of the theory of the thermal breakdown in solid (not porous!) dielectrics. This paper was followed by others [32–35] which widened and enriched our knowledge of physics of this phenomenon. The general theory of thermal breakdown was presented in the paper [36].

All these researches may be united by a single principle: they use two close approaches – the equation of heat conduction taking into account the heterogeneity of the electric field distribution in the volume, and the kinetic equation. Indeed, in order to obtain the equation of energy balance which takes into account both the heat transfer and the presence of an inhomogeneous electric field we should write the relation for the thermal energy dissipated in a time unit. We have

$$\frac{TdS}{dt} = \int \boldsymbol{j}\boldsymbol{E}\mathrm{d}v - \int q\mathrm{d}\boldsymbol{S} \geq 0,$$ (4.10)

where the heat flow $\boldsymbol{q} = -\kappa\nabla T$ and the current density $j = \sigma\boldsymbol{E}$. Taking $\boldsymbol{E} = -\nabla\varphi$, where φ is the field potential, we find using the Gauss theorem:

$$\int T\left(\frac{\mathrm{d}S_V}{\mathrm{d}t}\right)\mathrm{d}v = \int \nabla(\kappa\nabla T)\mathrm{d}v + \int(\sigma\nabla\varphi)^2\mathrm{d}v,$$

where S_V is the volume density of the entropy.

Since

$$T\left(\frac{\mathrm{d}S_V}{\mathrm{d}t}\right) = T\left(\frac{\partial S_V}{\partial T}\right)_P\left(\frac{\partial T}{\partial t}\right) = \frac{c_P\partial T}{\partial t},$$

where c_P is the isobaric heat capacity of a volume unit, the equation is

$$c_P\frac{\partial T}{\partial t} = \nabla(\kappa\nabla T) + \sigma(\nabla\varphi)^2. \tag{4.11}$$

In order to obtain a closed system of equations we need one more equation. Really, since the electromagnetic field energy is spent on heat, the whole dissipation may be presented in the following form:

$$\frac{\mathrm{d}U}{\mathrm{d}t} = \int \boldsymbol{E}\left(\frac{\mathrm{d}\boldsymbol{D}}{\mathrm{d}t}\right)\frac{\mathrm{d}v}{4\pi} - \int \nabla(\kappa\nabla T)\mathrm{d}v.$$

As $\boldsymbol{D} = \varepsilon\boldsymbol{E}$, where ε is the dielectric permeability, then putting this relation equal to zero we find the second equation:

$$\left(\frac{\varepsilon}{4\pi}\right)\boldsymbol{E}\frac{\mathrm{d}\boldsymbol{E}}{\mathrm{d}t} = \nabla(\kappa\nabla T). \tag{4.12}$$

The Eqs. (4.10) and (4.12) with definite initial and boundary conditions are "reference points" in developing the theory of thermal breakdown in solid dielectrics.

Solving these equations in various limiting cases we may describe the thermal breakdown of a medium in the nonstationary conditions of changing temperature and changing field.

In passing over to a porous medium the heat capacity c_P is considered to be an additive quantity, i.e. (see above)

$$c_P = (1 - m)c_{1P} + mc_{2P}, \tag{4.13}$$

where c_{1P} and c_{2P} are the heat capacities of the solid matrix and the gaseous phase, respectively. There are analogous relations for the temperature-conductivity coefficient, the resistance and the dielectric permeability of the structure [37–40]:

$$\begin{cases} \chi = (1 - m)\chi_1 + m\chi_2, \\ \rho = (1 - m)\rho_1 + m\rho_2, \\ \varepsilon = (1 - m)\varepsilon_1 + m\varepsilon_2. \end{cases} \tag{4.14}$$

Taking account of relations (4.14) and Eqs. (4.11) and (4.12) we may describe the distribution of the temperature and electric field in a porous medium. It should be noted that such a problem due to the nonlinear character of the equations is rather complicated and can be solved in particular practical cases only by fast-acting modern computers, though in a number of simple and practically uninteresting cases the system is solved easily.

As to the sphere of applicability of Eqs. (4.11) and (4.12), it lies in the validity of the condition imposed on the characteristic measurement range of temperature and electric field:

$$\delta x \gg R_1 - R_2 \,, \tag{4.15}$$

where $R_{1,2}$ are the external and internal radii of a fiber, respectively.

In the inverse limiting case it is necessary to account for the distribution of the fields and temperatures inside each area: both in the fibrils and the pores. It is a very difficult problem and need not be analyzed here, though its solution was outlined in papers [36,38]. It may seem that the condition (4.15) automatically "withdraws" from consideration the most interesting range of change of the parameters of concern to us. However, this is a seeming paradox. The thing is that the validity of inequality (4.15) absolutely does not affect the most important characteristic of the medium – its porosity. Really, the principal and the most interesting result manifests itself only when we take account of the dependence of all the values on the parameter m, but here the information about m is not lost in spite of the inequality (4.15)!

4.2 The Alternative Theory of Electric Breakdown in Porous Structures

The study of the electric breakdown of liquid and gaseous dielectrics started at the beginning of the 20th century and it was caused by the use of various isolation materials in electro-technical devices. The most intensive theoretical and experimental developments were carried out in 30–50 years of the 20th century [41–49]. Nevertheless, in spite of a great number of papers, two important subjects remained unclear and unrevealed in any known publication. We mean, first of all, the calculation of the breakdown current as a function of the intensity of the applied electric field E, and secondly, paradoxical though it may seem, the analytical dependence of the breakdown field on the structure parameters has not been determined.

The approach which we shall choose and substantiate for solution of these two problems is interesting, among other things, because it outlines methods to solve such problems and allows us to find any analytical dependencies of the parameters of our concern. Besides, these techniques allow us to bring all the mathematical calculations to their logical completion, which is very important from a purely theoretical point of view.

The following point of view on the process of the electric breakdown of a dielectric is commonly accepted and adequately founded. After passing the current of high intensity followed by the release of powerful thermal and sonic energy, there appears the non-zero stream of Joule heat directed inside the substance volume strictly normally to the direction of the electron stream. During its motion the electron stream "sweeps" everything on its path without deceleration in any local point of the volume: there does not exist such a cause which could somewhat seriously influence the effective energy dissipation of the electron beam moving with great velocity. Since the propagation time of this electron beam is very small and corresponds, according to experiments, in order of magnitude to approximately 10^{-7}–10^{-6} s, which corresponds to the time of electron scattering by phonons [25], consequently, on a hierarchic scale, it allows us to make a conclusion that the electron scattering *has no time* to occur and the role of "effective" scatterer must belong to some elastic mechanism. Such an elastic mechanism may be, for example, scattering of electrons by fluctuations of the dipole moments of atoms polarized by a strong external electric field. For the process of elastic scattering of electrons by fluctuations of the dipole moment only two directions of the scattering are of interest: one direction strictly forward and the other exactly backward. The backward-scattered electrons will collide with the oppositely moving electrons of the "breakdown current".

In the meeting of these two currents of opposite directions followed by internal friction a great quantity of heat will be released, which is just the thermal effect observed in practice. In microscopic terms it is the effect of the scattering of electrons by each other with the release (absorption) of photons. It is known [25] that the elastic mechanism of scattering is characterized by the highest probability of scattering per unit time (for this process the phase-space region is a maximum), and that is why in theoretical analysis in the framework of Boltzmann's kinetic equation only this mechanism should be taken into account. Note that the above reasoning does not contradict the commonly accepted statements of the theory of breakdown but only complement it [23, 41].

Before we pass over to a rigorous mathematical presentation of the theory of "elastic" breakdown we shall pay attention to four main statements:

1. The average electron energy in the breakdown current is not sufficient to ionize atoms in the fibril structure. This means that the following condition should be satisfied: $eE\langle a \rangle \ll Ze^2/\langle a \rangle$, where $\langle a \rangle$ is the mean "interatomic" distance, e is the electron charge, and Z is the number of electrons in the atom.

2. The elastic mechanism is the most effective scattering mechanism and allows to introduce some quite definite nonlinear conduction $\sigma(E)$ (see below).

3. The effectiveness of deceleration of the electron stream passing forward will be provided only by the scattering of the preceding electrons backward.

4. The electrons reflected backward in colliding with the oppositely moving electron stream cause the effective scattering of the beam in the plane $x - y$ (Fig. 4.1) and, as consequence, cause the thermal effect.

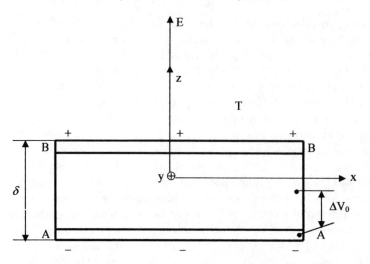

Fig. 4.1. Geometric pattern of the sample placed in a stationary electric field E. The field E is normal to the sample plane. T is the temperature and ΔV_0 is the barrier which electrons must overcome in order to get from the sample surface into its volume.

The stated problem may be solved by means of the method of Boltzmann's kinetic equation, which describes the dependence of the electron distribution function on the coordinates, momentum and time. This equation "works" in the quasi-classical approximation when the de Broglie electron wavelength is much greater than its free path l_e, i.e. $\lambda_b \gg l_e$. We shall divide the analysis of the problem into two parts: first, we shall consider only a solid dielectric, and then – after presentation of the main calculation technique – we shall pass over to the porous medium.

So, let us present Boltzmann's equation in the one-dimensional case in the following way:

$$\frac{\partial f}{\partial t} + eE\left(\frac{\partial f}{\partial p_z}\right) + v_z\left(\frac{\partial f}{\partial z}\right) = L\{f\}, \qquad (4.16)$$

where v_2 is the electron speed along the z axis and p_z is its momentum. The function

$$f = f(p_z, z, t) \qquad (4.17)$$

is the electron distribution function in the one-dimensional case when the dielectric thickness δ is large compared to the linear dimension of the breakdown area r_{br} $(r_{br} \ll \delta)$.

In order to find the integral of collisions $L\{f\}$ appearing on the right-hand side of (4.16) we shall write down the energy of interaction of an electron with the charge $Z_i e$, localized at the point with coordinate R_i in the following way:

$$V = e\varphi = \frac{Z_i e^2 \exp(-\lambda r)}{|r - R_i|},$$ (4.18)

and the analogous interaction with dipole moment fluctuations is

$$V_d = -E_i \delta d_i = e \delta d_i e^{-\lambda r} \nabla |r - R_i|^{-1},$$ (4.19)

where $\delta d = d - \langle d \rangle$, $\langle d \rangle$ is the mean equilibrium value of the dipole moment and λ is the inverse radius of the charge screening. Then, putting down the interaction (4.18) in representation of secondary quantization [54] and presenting the electron wave functions as a superposition of plane waves, after summation over all the dipole moments which are in the volume of the breakdown area $V_{br} = \pi_{br}^2 \delta$ we find that the interaction (4.18) takes the form:

$$V = \sum_{k,k'} \psi(k, k') c_k^+ c_{k'} + c.c.,$$ (4.20)

where $k = k_z$, $k' = k'_z$ and the scattering amplitude is

$$\psi(k, k') = \frac{e^2 \langle Z \rangle N}{V_{br}[(k - k')^2 + \lambda^2]},$$ (4.21)

where $\langle Z \rangle$ is the average quantity of electrons in an atom, $N = N_x N_y N_z$ is the number of polarized complexes in the volume V_{br}, $N_x N_y N_z$ is their number over axes x, y, z, and the letters $c.c.$ denote the complex conjugate.

In derivation of the relation (4.21) it was supposed that the distribution of dipole moments over the dielectric depth was uniform. The latter enabled us to replace the "charge" $Z_i e$ by $\langle Z \rangle e$. In case the distribution is not uniform, Gaussian for example, averaging should be made taking account of the dependence of Z_i on the coordinate z.

Thus, knowing the interaction energy (4.20), we can write the equation for the integral of collisions. Indeed, we have

$$L\{f\} = 2\pi \sum_{k'} \psi \left((p, p')(f_p - f'_p)\delta(\varepsilon(p) - \varepsilon(p'))\right),$$ (4.22)

where we take the law of electron dispersion as a square-law, i.e. $\varepsilon(p) = p_z^2/2m_e$, m_e is the electron mass, and the momentum p_z is connected with the wave vector k_z by Planck's relation $p_z = \hbar k_z$.

It should be noted that in the one-dimensional case Boltzmann's equation with the collision integral (4.22) can be solved precisely. It is important not so much for the practical aspect as for the methodical one. Moreover, Boltzmann's equation as the class of integral-differential nonlinear equations is solved in the so-called τ-approximation.

So, let us dwell upon the detailed solution of the given one-dimensional equation. We shall expand for this purpose the distribution function f into a Fourier integral over time and coordinates:

$$f(p, z, t) = \int\limits_{-\infty}^{\infty} \int\limits_{-\infty}^{\infty} e^{i\omega t - ikz} f_k(\omega, p) \frac{dk d\omega}{(2\pi)^2}.$$ (4.23)

Now we multiply both sides of (4.16) by pe^{iup} and integrate over p_z in the limits from $-\infty$ to $+\infty$. Then as a result of simple purely algebraic transformations and integration in parts we get the following differential equation for the distribution function transform f^*:

$$f^{*\prime\prime} + i(1 - Ax)f^{*\prime} - iA\left(\frac{1 - E^*\varphi(x)}{E}\right)f^* = 0\,, \tag{4.24}$$

where the new dimensionless parameters are

$$A = \frac{eEk}{m_e\omega^2}\,,$$

$$x = \frac{m_e\omega u}{k}$$

and the function

$$\varphi(x) = 0.5l\lambda^3 \left(\frac{\partial}{\partial\lambda}\right)\left\{\left[\frac{(\exp(-k\lambda|x|/m_e\omega))}{\lambda} - \frac{(\exp(-kl|x|/m_e\omega))}{l}\right]\right.$$

$$\left.\times(l^2 - \lambda^2)^{-1} - \frac{1}{\lambda}(1 + \lambda)\right\}\,. \tag{4.25}$$

Note that the artificially introduced by the equation

$$\delta(\varepsilon(p) - \varepsilon(p')) = v_z^{-1}\delta(p - p') = (\pi v_z)^{-1}(p - p' + i\hbar l)^{-1}\,.$$

The length of the electron free path l is a new parameter, ensuring the convergence of the solution. The field is given by

$$E^* = 8\pi \left(\frac{r_0^4\langle Z\rangle^2 e^3 m_e}{d_z^3 r_{br}^2 \hbar^2}\right)\,, \tag{4.26}$$

where the radius of screening is

$$r_0 = \frac{1}{\lambda}\,,$$

d_z is the mean distance between charges along the z axis and

$$l \ll d_z \ll \delta\,.$$

The function $\varphi(x)$ is even and continuous at the point $x = 0$ and its graphic form is given in Fig. 4.2.

The analysis of the Eq. (4.24) is difficult because the function $\varphi(x)$ is complicated. However, in the limit where λ and l tend to zero (or at x tending to infinity) this difficulty may be avoided. Really, as we shall see later, the greatest contribution to all the calculated parameters is made by the region of momentum near zero, that is why in solving (4.24) we can use the asymptotic behavior of the function $\varphi(x)$ at large x. It gives $\varphi(x) = 1$ and in this case we get quite a simple equation:

$$f^{*\prime\prime} + if^{*\prime}(1 - Ax) - iA\tau f^* = 0\,, \tag{4.27}$$

where $\tau = 1 - E^*/E$ and $E > E^*$.

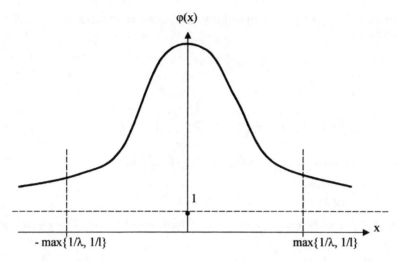

Fig. 4.2. Dependence of the function φ from the argument

By means of substitution we get

$$f^* = \exp\left(\frac{u^2}{4}\right) y(v),$$

where the new argument is

$$v = (2A)^{-1/2}(1 - Ax)(1 + \mathrm{i}).$$

Equation (4.27) is reduced to the equation of Weber–Hermit for the new function $y(v)$, and its solution having physical meaning may be presented so as

$$f^* = C \exp\left(\frac{y^2}{4}\right) D_{\tau-1}(\mathrm{i}v), \tag{4.28}$$

where $D_{\tau-1}(x)$ is the parabolic cylinder function and the constant C will be found later.

In order to find the transform of the function f^* we have to make an inverse Fourier transformation over the variables ω and u, and a Mellin transformation over the variable k. However, before starting this procedure let us pay attention to the fact that in calculating the breakdown current $j(E)$ it will be sufficient to take the stationary distribution function $f_{st} = f(p, z, \infty)$. It means that after the calculation of the integrals over k and u we should take $\omega = 0$ and omit integration with respect to frequency. Such a procedure greatly simplifies the calculations and therefore, if we use the integral representation for the parabolic cylinder function (see [55]), we obtain

$$D_{\tau-1}(\mathrm{i}v) = \left[\frac{\exp(y^2/4)}{\Gamma(1-\tau)}\right] \int\limits_{0}^{\infty} \exp\left(-\mathrm{i}v\xi - \frac{\xi^2}{2}\right) \xi^{-\tau} \mathrm{d}\xi.$$

The formula for the sought-for distribution function in the stationary case will take the form

$$f_{sc} = \left[\frac{C}{\Gamma(1-\tau)}\right] \int_{-\infty}^{\infty} e^{-ipu}du \tag{4.29}$$

$$\times \lim_{\omega \to 0} \int_{0}^{\infty} e^{-ikz}dk \int_{0}^{\infty} \exp\left(\frac{v^2}{2} - iv\xi - \frac{\xi^2}{2}\right) \xi^{-\tau}d\xi,$$

where we take account of the explicit form of the argument

$$v = (1+\mathrm{i})\frac{(1-eEu/\omega)}{(2A)^{1/2}}.$$

Equation (4.29) may be easily calculated. Indeed, using Fresnel's formulae

$$J = \int \exp\{ix^2\}dx = \left(\frac{\pi}{2}\right)^{1/2}(1\pm\mathrm{i}),$$

we finally find

$$f_{sc} = \frac{C(2\pi)^{1/2}\Gamma(1+\tau/2)e^{-i\pi\tau/2}p^{\tau-1}(m_e eE)^{-(1-\tau)/2}}{(z+p^2/2m_e eE)^{1+\tau/2}}. \tag{4.30}$$

Now in order to estimate the constant C we use the normalization condition

$$\int_{0}^{\infty}\int_{d_z}^{\delta} f_{sc}(p,z,\infty)\frac{dpdz}{2\pi\hbar} = 1.$$

After simple calculation of the integral we obtain

$$C = \frac{\tau(2\pi)^{1/2}e^{i\pi\tau/2}\hbar}{2^{1+\tau/2}(em_e E)^{1/2}\Gamma(1+\tau/2)\ln(\delta/d_z)}. \tag{4.31}$$

Using Eqs. (4.30) and (4.31) it is easy to calculate the breakdown current as a function of the electric field intensity E at $E \gg E^*$. Indeed, according to the definition, the current is

$$j(E) = \left(\frac{e}{2\pi\hbar m_e}\right)V_{br}\int_{0}^{\delta} dz \int_{0}^{\infty} pf_{sc}(p,z,\infty)dp.$$

Substituting here (4.30) and including the relation (4.31) we get

$$j(E) = \frac{\tau^2\Gamma(\tau)(e^3 E\delta/m_e)^{1/2}}{2^\tau\pi\Gamma^2(1+\tau/2)\ln(\delta/d_z)}. \tag{4.32}$$

Hence, in particular it is clear that at $E \Rightarrow E^* + \mathrm{i}0$ the asymptotic behavior of current is

$$j(E) \Rightarrow \Gamma(\tau) \cong \ln\left(\frac{1-E^*}{E}\right). \tag{4.33}$$

j(E)

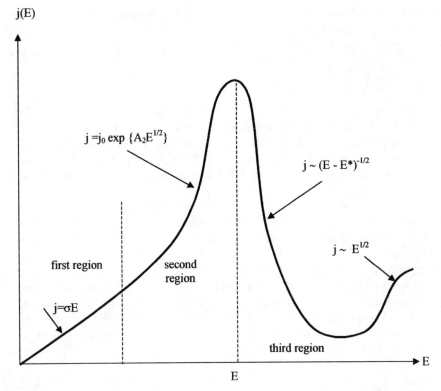

Fig. 4.3. Schematic representation dependence of the breakdown current j on the intensity of the external electric field E. Region 1 corresponds to Ohm's law; region 2 corresponds to Frenkel's law; region 3 is the dependence described in the text.

A graphic representation of the current density as a function of the external electric field intensity E is illustrated in Fig. 4.3. As we may see from relation (4.32) the breakdown current is not described by Ohm's differential law and has a clearly nonlinear character. In this respect it should be noted that in the equation for the free energy F in the region of fields close to the field E^* a logarithmic singularity will arise (through the current $j(E)$ leading to the singularity of E and all its derivatives with respect to E). In the range of values $E \gg E^*$ the current j should have behaved as the function $E^{1/2}$, but due to the inaccessibility of such values for experimentation it is no use discussing it here.

In order to estimate the numerical value of the breakdown current let us assume the following parameters included in the formula (4.32). Indeed, let, for example, $\delta = 10^{-2}$ cm, $r_{br} = 10^{-5}$ cm, $m_e = 10^{-27}$ g, $e = 4.8 \times 10^{-10}$ (CGS), $\tau = 0.5$, $E = 6 \times 10^6$ V/cm $= 2 \times 10^3$ (CGS), and suppose that the field of breakdown $E^* = 10^3$ (CGS). Then we immediately obtain for the critical current that $j_{br} = 10A/cm^2$.

Let us estimate the breakdown field according to (4.26). As

$$E^* = \frac{8\pi r_0^4 \langle Z \rangle e^3 m_e}{d_z^3 r_{br}^2 \hbar^2},$$

then taking here $\langle Z \rangle = 30$, $r_{br} = 10^{-6}$ cm (the latter assumption that $r_0 = 0.1 d_z$ is caused by the quite natural requirement that the radius of screening should not be larger than the distances between complexes – dipole moments), we obtain the result $E^* \cong 10^6$ V/cm. Such an estimate is in satisfactory agreement with a large number of experiments (see the classic monograph [23]).

Let us see how we can modify the formula (4.26) in the case of electron scattering caused by fluctuations of the dipole moments. Really, as the value $e\langle Z \rangle \langle a \rangle$ is approximately equal to the dipole moment of the atom $|d|$, then replacing the value $e\langle Z \rangle \langle a \rangle$ by $\langle \delta d^2 \rangle$, where the angle brackets mean averaging over fluctuations of the dipole moments of atoms in the region of the breakdown zone, and taking into account on the basis of the general theory of fluctuation phenomena that $\langle \delta d^2 \rangle = T v_{br}/N = T b^3$ (see [11]), we get the modified dependence of the breakdown field for the case when the main role belongs to the mechanism of electron elastic scattering by fluctuations of the dipole moment directions:

$$E_d = \frac{8\pi r_0^4 \langle Z \rangle e T m_e}{d_z^3 r_{br}^2 \hbar^2 \langle a \rangle^2}. \tag{4.34}$$

We did not focus our attention on the fact that the breakdown field must depend not only on the classic constants (electron charge, Planck's constant, etc.) but also on the purely geometric characteristics of the structure. This very important point plays a significant role in understanding the very definition of the electric breakdown field because, as numerous experiments show [57–66], it is defined in most cases only by the purely geometric parameters of the structure. The other thing is that without electron microscopy it is rather difficult to evaluate the influence of the various geometric factors on the value E^*, which, of course, would be desirable to do before the beginning of every experimental study of the electric breakdown field.

For the chaotic distribution of pores about the structure volume the question of how to describe the dependence of E^* on the additional characteristic – the porosity m – has a somewhat philosophical character. This is, firstly, due to the fact that it is not known how many pores are in the local area between the points of application of the external potential difference.

Secondly, it is not clear how to count them. These two factors first of all indicate that the description of the breakdown field in such substances must have a purely probability character. The formula (4.9) may satisfactorily describe the breakdown field in the case of alternating, like a sandwich, regions.

Such structures, as we know [67], are called composite materials, but in the present monograph these types of "compounds" are not considered.

Thus, in order to find out how it is possible to account for the substance porosity m and to represent the breakdown field in the form of a concrete formula we should remember that during the motion of the highly energetic electron stream though the structure it must (according to the hierarchic principle) be scattered elastically by any defects, including fluctuating dipole moments localized both in the volume of the pores and in the solid matrix. Hence, the common contribution to E^* from the two different streams must have (at least to the first approximation in porosity) a purely additive character similar to Matissen's rule in metals [24, 25].

The breakdown field in gas for the elastic mechanism of electron scattering by gas-phase molecules may be written in the form

$$E_g^* = \frac{2\pi T m c_i \sigma(T)}{e},$$

where c_i is the molecule concentration in a unit volume and $\sigma(T)$ is the cross-section of electron elastic scattering by molecules (cf. [24]).

In order to account for the probability of the presence of W in the pore beam path, E_g^* should be multiplied by W. Thus the sought-for breakdown field for a porous medium is

$$E_p^* = 2\pi T \left\{ \frac{W m c_i^\sigma(T)}{e} + \frac{4(1-m)\pi r_0^4 \langle Z \rangle e m_e}{d_z^3 r_{br}^2 \hbar^2 \langle a \rangle^2} \right\}. \tag{4.35}$$

From the given relation, in particular, we can see that in the range of relatively high temperatures the breakdown field increases linearly with T. The question arises of how to explain the experimental data, according to which the breakdown field of a solid dielectric at first increases (with the temperature increase) and then falls (Fig. 4.4). Let us consider the qualitative aspect of this phenomenon. Indeed, in the range of low temperatures, lower than some (so far purely abstract!) temperature T_0, there is only one electron stream which appears as result of ionization of atoms in the dielectric internal structure by a strong stationary electric field. This avalanche stream self-accelerating in the field E and impelling into motion more and more electrons in its path considerably damages the structure, which results, first of all, in the thermal effect. With the temperature increase the situation begins gradually to change and at some moment (on the temperature scale) two electron streams must appear: one stream is the same as at $T < T_0$, and the other appears due to the increased probability of tunnel penetration of electrons (streaming off from the cathode on the dielectric surface) into the structure volume. The corresponding probability is described by the Arrhenius dependence: $W_a = \nu \exp\{-\varepsilon_a/T\}$, where ε_a is the energy barrier overcome by electrons in the tunneling process and ν is the characteristic electron frequency. If there are two streams which must cause the same dielectric damage in the temperature range $T > T_0$ as at $T < T_0$, when there is

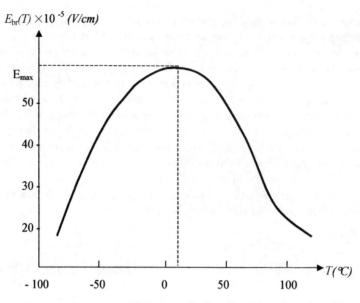

Fig. 4.4. Symbolic pattern of the breakdown field dependence on the temperature T

only one stream, it is necessary to decrease their combined energy. The latter may be achieved only by decreasing the field E_m^*.

Let us now try to express this qualitative physical reasoning in rigorous mathematical terms.

We shall choose the system of coordinates as shown in Fig. 4.1. The external electric field E will be directed along the z axis. According to the general rules of quantum mechanics [55] the wave function of electrons must be expanded in terms of a superposition of eigenfunctions characteristic of a charged particle in the stationary electric field E. The role of such functions is performed by Airy functions. That is why the wave function

$$\psi(\boldsymbol{x}) = N_0^{-1/2} \sum_{\{p\}} c_p Ai(z) \exp(ip_x^x + ip_y^y) \,, \tag{4.36a}$$

where c_p^+ is the annihilation operator of an electron with momentum $\boldsymbol{p}$. An analogous relation holds for the complex-conjugate wave function as well:

$$\psi^+(\boldsymbol{x}) = N_0^{-1/2} \sum_{\{p\}} c_p^+ Ai(z) \exp\{-ip_x x - ip_y y\} \,, \tag{4.36b}$$

where c_p^+ is the creation operator of an electron with momentum $\boldsymbol{p}$ and the Airy function is given by

$$Ai(\xi) = B \int\limits_0^\infty \cos\left(u\xi + \frac{u3}{3}\right) du \,, \tag{4.37}$$

where

$$\xi = \left(\frac{z+\varepsilon}{eE}\right)\left(\frac{2m_e eE}{\hbar^2}\right)^{1/3}, \tag{4.38}$$

ε is the electron energy, and the constant B is given by

$$B = \frac{(2m_e)^{1/3}}{\pi^{1/2}(eE)^{1/6}\hbar^{2/3}V_{br}^{1/2}}. \tag{4.39}$$

In the beginning of the present section we regarded the electron as a plane wave without taking into account the presence of the field E. It was correct only if the condition $\varepsilon \ll eE\langle a\rangle$ was valid, which was implied without explanation.

Now we shall deal with the general case when to take account of the field E is necessary and the electron kinetic energy ε cannot be neglected.

Let us write down the interaction of the electron with the guided dipole moment of atoms located in the volume V_{br} in the following form [41]:

$$V = -N_0^{-1}\sum_{i=1}^{N_0}\boldsymbol{E}_i\delta\boldsymbol{d}_i = N_0^{-1}\sum_{i=1}^{N_0}\delta\boldsymbol{d}_i\nabla\left(\frac{eZ_i}{|\boldsymbol{r}-\boldsymbol{R}_i|}\right), \tag{4.40}$$

where Z_i is the atomic number of the ith-atom.

After transition into the representation of secondary quantization, which is done by the substitution

$$V_{\text{int}} = \int \psi^*(\boldsymbol{x})V\psi(\boldsymbol{x})\mathrm{d}^3x, \tag{4.41}$$

where the function $\psi(\boldsymbol{x})$ is given by the Eqs. (4.36a, 4.36b) and taking into account the summation rule of Ewald [12] we obtain the equation of interest, connected with the mechanism of electron elastic scattering by dipole moment fluctuations of atoms. Indeed, as the asymptotic form of the Airy function at large positive z is

$$Ai(\xi) = 0.5B\pi^{1/2}\xi^{-1/4}\exp\left\{-\frac{2\xi^{3/2}}{3}\right\}, \tag{4.42}$$

then from the above we get

$$V_{\text{int}} = \Omega_0 \sum_{\{p,p'\}} c_p^+ c_{p'}, \tag{4.43}$$

where the process amplitude is

$$\Omega_0 = \frac{(2\pi)^2 2^{1/2}\delta d_z e\hbar\exp\{-E_0/E\}}{N_0(m_e\varepsilon)^{1/2}\langle a\rangle^3}, \tag{4.44}$$

$$E_0 = \frac{4\varepsilon^{3/2}(2m_e)^{1/2}}{3e\hbar}. \tag{4.45}$$

Remember that the complete number of atoms in the volume V_{br} is

$$N_0 = \frac{(\pi r_{br}^2 \delta)}{\langle a \rangle^3} . \tag{4.46a}$$

As we are dealing with a porous medium we should take

$$N_0 = \frac{(\pi r_{br}^2 \delta (1-m))}{\langle a \rangle^3} . \tag{4.46b}$$

Similarly to the Eq. (4.27) we can obtain for the function of the electron distribution over the coordinate and momentum space in the case of interactions (4.43) the following simple equation

$$f'' + i f'(1 - A^* x) - i A^* \tau f = 0 , \tag{4.47}$$

where the original of the distribution function is $f(p, z, t) = \langle c_p^+ c_p' \rangle$ and the angle brackets characterize the averaging over the ground state of the electron subsystem. Here it is the state at absolute zero temperature.

The parameter τ is given by

$$\tau^* = \frac{1 - E^{**}}{E} ,$$

and unlike the field E^* the field E^{**} is given by

$$E^{**} = \frac{2 \Omega_0^2 m_e \langle a \rangle}{e \hbar^2} .$$

In detailed form taking account of (4.44) we have

$$E^* = \frac{4(4\pi)^4 e \langle a \rangle \langle \delta d^2 \rangle \exp\{-2E_0/E\}}{\varepsilon V_{br}^2} .$$

Since $\langle \delta d^2 \rangle = T V_{br}/N_0$ and N_0 is given by the Eq. (4.46a), we finally find

$$E^* = \frac{1024 \pi^2 e T \langle a \rangle^4 \exp\{-2E_0/E\}}{\varepsilon r_{br}^4 (1-m)^3 \delta^2} . \tag{4.48}$$

In the region of low-frequency values, when $E > 2E_0$, the formula (4.48) at $T = 3 \times 10^{-14}$ erg, $\delta = 10^3$ cm, $r_{br} = 3 \times 10^{-6}$ cm, $\langle a \rangle = 3 \times 10^{-7}$ cm, $e = 4.8 \times 10^{-10}$ CGS, $\hbar = 10^{-27}$ erg s, gives

$$E^{**} \cong 3 \times 10^6 \, \text{V/cm} .$$

The present estimate is quite consistent with the experimental results [58–60]. Note, by the way, that the qualitative dependence of the breakdown field E^{**} on the dielectric thickness, the dielectric behaving as δ^{-2}, also agrees with the experimental results not only of the present day, but with those described previously, e.g. in [23].

Now we shall analyze the temperature dependence E^{**}. If the condition $E > 2E_0$ holds, then at values of electron energy ε the breakdown field linearly rises with the temperature. At the temperature increase the electron kinetic energy rises and then it turns out that $E < 2E_0$, but here E^{**} will be proportional to $e^{-T^{3/2}}$. The latter, in turn, shows that the breakdown field must decrease with the temperature increase. The plot of E^{**} versus temperature is given in Fig. 4.4.

Let us estimate the temperature at which E^{**} is a maximum. In reality E^{**} is given by

$$E^{**} = D_1 T \exp\{-D_2 T^{3/2}\},$$

where the constants $D_{1,2}$ are defined by the equations

$$D_1 = \frac{1024\pi^2 e \langle a \rangle^4}{\varepsilon r_{br}^4 (1 - m)^3 \delta^2},$$

$$D_2 = \frac{8(2m_e)^{1/2}}{3eE\hbar}.$$

Differentiating the function E^{**} with respect to temperature and solving the obtained equation we find that in order of magnitude

$$T_0 = \frac{8(eE\hbar)^{2/3}}{m_e^{1/3}}. \tag{4.49}$$

The obtained expression for T_0 corresponds nearly to 270 K.

We shall underline that the formula (4.48) has a universal character and allows us to describe the different parametric dependence of the electric breakdown field from the dielectric thickness and temperature. Now we shall try to answer the question about the possibility of application of the f-points model in order to describe the breakdown field of the porous fibril structure [68–71]. To answer this question we shall preliminarily suppose that the electron wavelength λ complies with the inequality

$$\lambda_B \ll l_{fl} \ll \lambda \ll l, \tag{4.50}$$

where λ_B is the electron wavelength of de Broglie, l is the free path length and l_{fl} is the fibril size. Let us consider two cases of practical interest.

$$1. \qquad l_M \ll l_{fl} \ll \lambda \ll l \ll D, \tag{4.51}$$

here l_M is the length of a molecular free path in the gaseous phase (in pores) and D is the distance between fibrils (see Fig. 3.3b for $D_1 = D_2 = D_3 = D$). The given inequality implies that the dielectric structure is very porous and hence the condition

$$1 - m \ll 1$$

is valid.

Due to the inequality (4.51) we should believe that the electron scattering by every fibril segment occurs independently from each other. Because $l_m \ll \lambda \ll D$, the loss of energy by the electron beam in the gaseous phase may be calculated using the gas dynamics equations. In order to understand how to solve the stated problem, we shall write the equation describing the loss of average kinetic energy in the electron stream as:

$$\frac{d\langle\varepsilon\rangle}{dt} = -\gamma\langle\varepsilon\rangle,\tag{4.52}$$

where the sought-for coefficient of losses is

$$\gamma = m\gamma_M + (1-m)\gamma_{fl},\tag{4.53}$$

here γ_{fl} is the probability of electron scattering per unit time by the fibrils and γ_M is the probability of its scattering in the gaseous phase by density fluctuations. Let us clarify how the electric breakdown field is connected with the coefficient γ. For it we shall write the motion equation for the average speed of the electron stream:

$$\frac{m_e d\langle v_e\rangle}{dt} = eE.$$

Hence in order of magnitude

$$\langle v_e\rangle = \frac{eE\tau}{m_e},$$

and therefore

$$\tau^{-1} = \frac{eE}{m_e}\langle v_e\rangle.\tag{4.54}$$

Then, since

$$\langle v_e\rangle = \frac{\delta}{\tau},$$

from the relation (4.54) it immediately follows that

$$\tau^{-1} = \left(\frac{eE}{m_e\delta}\right)^{1/2}.\tag{4.55}$$

Assuming that electric breakdown occurs if the following condition is valid:

$$\tau^{-1} \geq \gamma,\tag{4.56}$$

we may by taking account of Eqs. (4.53) and (4.55) find the minimum value $E_{\min} = E_{br}$ at which the given inequality holds.

The result is

$$E_{br} = \frac{[(1-m)\gamma_{fl} + m\gamma_M]^2 m_e\delta}{e}.\tag{4.57}$$

Now the task is to calculate the attenuation γ_M and γ_{fl}. Let us start from γ_M. As for the conditions in question the electron stream may be regarded as a "classic liquid". The average speed of this stream must comply with the equation of Navies–Stokes, which in this case accounts for the deceleration by the gaseous medium with viscosity η_g:

$$\rho_e \frac{\partial \langle v \rangle}{\partial t} = \eta_g \Delta \langle v \rangle \,, \tag{4.58}$$

where $\rho_e = m_e n_e$ is the electron beam density and n_e is the electron concentration per unit volume.

As characteristic changes of the average speed $\langle v \rangle$ occur at distances of the order of the linear dimensions of the breakdown region, e.g. δ, Eq. (4.58) will be rewritten in the following way

$$\frac{d \langle v \rangle}{dt} = -\gamma_M \langle v \rangle \,, \tag{4.59}$$

where the sought-for coefficient of "losses" is

$$\gamma_M = \frac{\eta_g}{\rho_e \delta^2} \,. \tag{4.60}$$

Now, in order to calculate the attenuation γ_F, imagine that the "electron liquid" attacking the fibrils is at rest, while the fibrils move in the opposite direction with velocity $\langle v \rangle$. It means that every fibril is subject to Stokes deceleration and its motion in such a viscous medium may be estimated using the equation:

$$\frac{d \langle v \rangle}{dt} = -\frac{k \eta_e l_{fl} \langle v \rangle}{m_{fl}} \,, \tag{4.61}$$

where the coefficient k accounts for the non-spherical character of the fibrils (for a sphere $k = 6\pi$), $\eta_e = \rho_e \nu_e$ is the dynamic viscosity of the electron beam, ν_e is the kinematic viscosity and m_{fl} is the mass of the fibrils.

From (4.61), assuming that in the electron path there may be N_{fl} fibrils, it immediately follows that

$$\gamma_{fl} = \frac{k \eta_e l_{fl}}{m_{fl} N_{fl}} \,. \tag{4.62}$$

Since

$$m_{fl} = \rho_{fl} v_{fl} = \rho_{fl} \pi \langle r_{fl} \rangle^2 l_{fl} \tag{4.62a}$$

and

$$N_F = \frac{v_{fl}}{\pi \langle r_{fl} \rangle^2 l_{fl}} = (1 - m) \frac{\pi r_{br}^2 \delta}{\pi \langle r_{fl} \rangle^2 l_{fl}} = (1 - m) \left(\frac{r_{br}}{\langle r_{fl} \langle} \right)^2 \left(\frac{\delta}{l_{fl}} \right) \,,$$

we finally get

$$\gamma_{fl} = \frac{k \eta_e l_{fl}}{\pi \rho_{fl}} (1 - m) r_{br}^2 \delta \,. \tag{4.63}$$

Substituting now (4.60) and (4.63) into (4.57) and replacing $m_e \Rightarrow \pi r_{br}^2 \rho_e \delta$ we shall find for the breakdown field the expression sought-for:

$$E_{br} = \left(\frac{k \eta_e l_{fl}}{\pi \rho_{fl} r_{br}^2} + m \eta_g \rho_e \delta \right)^2 \left(\frac{\rho_e r_{br}^2}{e} \right) \,. \tag{4.64}$$

The formula (4.64) allows us to give practical recommendations which can facilitate the correct choice of porous structures with the best electric strength, which has already been rigorously proved theoretically.

Thus, such structures should have:

1. maximum dimensions of fibrils;
2. minimum thickness δ;
3. minimum dimensions of inhomogeneity of the surface, r_{br};
4. maximum viscosity of fluids saturating the porous medium, η_g.

Anticipating the obvious question of why we need so many different points of view on the quantitative estimations of the breakdown field, we shall answer that the necessity to analyze this complicated phenomenon from different theoretical positions allows us:

1. to predict and then to verify experimentally the dependence of E_{br} (see also formulae for E^* and E^{**}) on every *new* parameter of the structure (dielectric thickness δ, porosity m, fibril dimension l_f, and so on),
2. to compare every new dependence with the earlier obtained ones in order to assure ourselves that they are not inconsistent and that their behavior is qualitatively correct.

The latter point, to tell the truth, has a more theoretical importance. Now we pass over to the second case:

$$2. \qquad l_M \ll \lambda \ll l \ll l_{fl} \ll D. \tag{4.65}$$

It is clearly seen that the present situation differs from the previous one in only one point: if the given inequality is valid, there can occur additional energy losses of the stream of the "electron liquid" (EL), connected with the process of thermal conduction. We shall clarify this point.

As a result of electron collisions with each other they acquire their own temperature T_e, which is equal on average to the temperature EL. Due to the possibility of the condition $l \ll l_{fl}$, the stream attacking the fibrils (whose temperature is T) begins gradually to give up its own thermal energy. This fact just shows that the process of thermal conduction takes place. Here the question arises of whether the electron stream will have time to "jump" from pole to pole without giving up its heat. To give an answer we should estimate the heating time of the fibrils. According to [75] we have

$$t_{hc} = \frac{l_{fl}^2}{\chi_{fl}}. \tag{4.66}$$

Taking here $l_{fl} = 10^{-4}$ cm and $\chi_{fl} = 10^2$ cm^2/sec (which, by the way, corresponds to the thermal diffusion of the solid body), we immediately obtain $t_{hc} = 10^{-10}$ sec. The time of breakdown, as it was pointed above, corresponds approximately to 10^{-6}–10^{-8} s.

Thus, relying upon the numerical estimation we may state that due to thermal conduction the electron stream partially gives its energy to the fibril structure. In order to account for the effect of thermal conduction in the formula (4.57) a simple formal substitution should be done:

$$\gamma_{fl} \Rightarrow t_{tc}^{-1},$$

where according to (4.66), taking account of the entire number of fibrils,

$$t_{tc}^{-1} = \frac{\chi_{fl}}{l_{fl}^2} N_{fl}$$

or, due to the relation (4.62a),

$$t_{tc}^{-1} = \left(\frac{\chi_{fl}}{(1-m)l_{fl}\delta} \right) \left(\frac{\langle r_{fl} \rangle}{r_{br}} \right)^2 .$$

Therefore, instead of (4.64) the breakdown field should be represented as follows:

$$E_{br} = \left[\left(\frac{\chi_{fl}}{l_{fl}} \right) \left(\frac{\langle r_{fl} \rangle}{r_{br}} \right)^2 + m\eta_g \rho_e \right]^2 \left(\frac{\rho_e r_{br}^2}{e} \right) . \tag{4.67}$$

Concluding the present section we should add that the above stated theory of electric breakdown has an alternative character (as we already mentioned earlier, the peak of the research into this issue may be referred to the 1950s) and, without touching upon the recognized priority papers, has only one purpose: to show the calculation technique for the breakdown field, calculating it by means of the one-dimensional Boltzmann equation (which is itself of purely theoretical interest) as a function of the various microscopic and macroscopic parameters of the structure.

4.3 The Dependence of the Breakdown Field on the Frequency of the External Oscillating Electric Field

Let us now assume that a porous dielectric is put into an external oscillating electric field $\boldsymbol{E} = \boldsymbol{E}_0 \cos \omega t$, where $\boldsymbol{E}_0$ is the amplitude and ω is the frequency. Here quite a reasonable question arises: how does the high-frequency application influence the magnitude of the breakdown field and how does this case differ from the case where the dielectric is in a stationary electric field. Let us first turn to experiment. Measurements of the breakdown showed that it falls with frequency at the rate of about $\omega^{-1/3}$ (see Fig. 4.5, dotted line). Qualitatively this result is clear: with the increase of frequency ω the electric "resistance" must decrease, and, consequently, the breakdown field will be smaller too. We shall demonstrate this. Indeed, as in the alternating field case, the breakdown current is $j = \sigma(\omega) = E^*(\omega)$, then on the assumption that a sample will be damaged to the same extent as in a stationary field, we may write the equation:

$$j = \sigma(\omega)E^*(\omega) = \sigma E^*$$

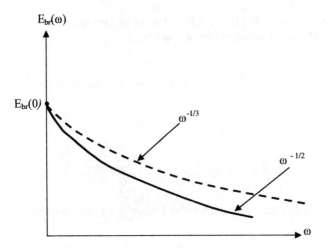

Fig. 4.5. Breakdown field dependence on the frequency of the external alternating electric field. Solid line is theory; dotted line is experiment.

and hence

$$E^*(\omega) = \frac{\sigma E^*}{\sigma(\omega)}, \qquad (4.68)$$

where E^* is defined by the formulae (4.26), (4.34), (4.35), (4.48), (4.64) and (4.67).

As we can see, for the calculation of the parametrical dependence $E^*(\omega)$ we should find the conductivity $\sigma(\omega)$. The conductivity $\sigma(\omega)$ according to the modified (for our case) Kubo's formula (for example, see the original papers [76] and [40]) may be presented as:

$$\sigma(\omega) = \sigma + \Delta\sigma \left\langle\!\!\left\langle \frac{\omega\tau_e(v_{fl})}{[1 + \omega^2\tau_e^2(v_{fl})]} \right\rangle\!\!\right\rangle, \qquad (4.69)$$

where σ is the static conductivity of the dielectric at $\omega = 0$, $\Delta\sigma$ is the value with the dimensions of conduction, and the double angle brackets mean that averaging is made twice: one time over the electron ensemble and the other time over the fibril dimension. $\tau_\varepsilon(v_{fl})$ is the relaxation time of the electrons during their scattering by a fibril with the volume v_{fl}.

Within such a setting of the problem a simple and obvious assumption suggests itself. Namely, we shall assume that the electron wavelength, $\hbar/p$, where p is its momentum, exceeds the fibril diameter:

$$2pr_{fl} \ll \hbar. \qquad (4.70)$$

Or, as $p = 2\pi\hbar/\lambda$, we have

$$\lambda \gg 4\pi r_{fl}. \qquad (4.71)$$

Now we shall try to estimate the probability of this wave scattering by one specified fibril. For this purpose it should be remembered that the scattering cross-section is

$$\Delta = \frac{W}{j} \,, \tag{4.72}$$

where $W = 1/\tau_\varepsilon(\omega_{fl})$ is the sought-for probability of the electron scattering per unit time and $j = n_e v_e = n_e p/m_e$, where n_e is the number of electrons in one cubic centimeter.

On the other hand, since the scattering cross-section may be represented in "Rayleighian" form (taking $\lambda \gg r_{fl}$) by

$$\Delta = \left(\frac{p}{\hbar}\right)^4 v_{fl}^2 \,, \tag{4.73}$$

then, according to (4.72), we find

$$\frac{1}{\tau_\varepsilon}(v_{fl}) = \frac{n_e p^5 v_{fl}^2}{m_e \hbar^4} = \frac{2^{5/2} n_e \varepsilon^{5/2} m_e^{3/2} v_{fl}^2}{\hbar^4} \,. \tag{4.74}$$

To find the dependence $\sigma(\omega)$ we should perform double averaging in (4.69). For this purpose we rewrite it in the more detailed form:

$$\sigma(\omega) = \sigma + \Delta\sigma Z^{-1} \int \int \langle \varphi(p) \rangle \left\{ \frac{\omega \tau_p(v)}{[1 + \omega^2 \tau_p^2(v)]} \right\} \frac{d^3 p \, dV_{br}}{(2\pi\hbar)^3} \,, \tag{4.75}$$

where Z is a normalization factor. The equilibrium function of the electron distribution will be chosen in the form of the classic Maxwell's distribution:

$$\langle \varphi(p) \rangle = \exp\left\{ \frac{-p^2}{2 m_e \varepsilon_0} \right\} \,, \tag{4.76}$$

where ε_0 is the average energy of the electron beam (in a particular case it may be equal to the temperature of the electrons at small times of their internal collisions).

As to the cell of the volume dV_{br}, it may be expressed through the function of fibril distribution over dimensions $f(v)$. Indeed, as $v_{br} = \pi r_{br}^2 \delta$, the number of fibrils is

$$N_{fl} = \frac{\pi r_{br}^2 (1-m)\delta}{\int f(v) dv} \,.$$

And therefore

$$dV_{br} = \frac{N_{fl} f(v) dv}{(1-m)} \,. \tag{4.77}$$

Substituting (4.76) and (4.77) into (4.75) we get

$$\sigma(\omega) = \sigma + \frac{N_{fl}\Delta\sigma}{2\pi^2 \hbar^3 Z(1-m)}$$

$$\times \int_0^\infty \int_0^\infty \exp\left(-\frac{p^2}{2m_e \varepsilon_0}\right) \frac{\omega \tau_p(v) f(v) dv p^2 dp}{1 + \omega^2 \tau_p^2(v)} \,. \tag{4.78}$$

Passing finally in this equation from integration over momentum to integration over energy, using the relation $\varepsilon = p^2/2m_e$, we represent the formula (4.78) in the form:

$$\sigma(\omega) = \sigma + \frac{2^{1/2} N_{fl} m_e^{3/2} \Delta\sigma}{\pi^2 \hbar^3 Z(1-m)} \int_0^\infty \exp\left(-\frac{\varepsilon}{\varepsilon_0}\right) \varepsilon^{3/2} d\varepsilon \int_0^\infty \frac{\omega \tau_p(v) f(v) dv}{1 + \omega^2 \tau_p^2(v)} . \quad (4.79)$$

The internal integral of (4.79) may be evaluated in a general form if the distribution function $f(v)$ is even. Really, then the integration limits may be extended from $-\infty$ up to $+\infty$ and then, using the Cauchy formula and the Jordan lemma [78] and by means of the residue theory, the integral value may be found:

$$J_{in} = \frac{\pi \omega \tau_\varepsilon(v^*) f(v^*)}{2 d(\omega^2 \tau_\varepsilon^2(v))/dv}\bigg|_{v=v^*} .$$

Let us assume a uniform distribution of the fibrils over the dimensions, i.e.

$$f(v) = \begin{cases} 1 \text{ at } v < \langle v_{fl} \rangle \\ 0 \text{ at } v > \langle v_{fl} \rangle \end{cases} . \quad (4.80)$$

Then, taking account of (4.74) we get

$$J_{in} = \int_0^\infty \frac{\omega \tau_p(v) f(v) dv}{1 + \omega^2 \tau_p^2(v)} = \int_0^\infty \frac{\alpha \omega v^2 dv}{v^4 + \alpha^2 \omega^2}$$

$$= (\alpha\omega)^{1/2} \int_0^\infty \frac{x^2 dx}{(1+x^4)} = \pi \left(\frac{\alpha\omega}{2}\right)^{1/2} , \quad (4.81)$$

where the parameter α is given by

$$\alpha = \frac{\hbar^4}{2^{5/2} n_e \varepsilon^{5/2} m_e^{3/2}} .$$

Taking account of the inner integral (4.81), the formula (4.79) may be rewritten in the following way:

$$\sigma(\omega) = \sigma(\omega) = \sigma + \frac{N_{fl} m_e^{3/4} \Delta\sigma \omega^{1/2}}{2^{5/4} \pi \hbar Z(1-m) n_e^{1/2}} \int_0^\infty \exp\left(-\frac{\varepsilon}{\varepsilon_0}\right) \varepsilon^{1/4} d\varepsilon .$$

The integral contained here is

$$J = \int_0^\infty \exp\left(-\frac{\varepsilon}{\varepsilon_0}\right) \varepsilon^{1/4} d\varepsilon = 0.25 \varepsilon_0^{1/4} \Gamma\left(\frac{1}{4}\right) ,$$

where $\Gamma(x)$ is the gamma function.

From the normalization condition we find

$$Z = \int_0^\infty \exp\left(-\frac{\varepsilon}{\varepsilon_0}\right) d\varepsilon = \varepsilon_0 . \tag{4.82}$$

And ultimately we have

$$\sigma(\omega) = \sigma + \Delta\sigma(\omega\langle\tau\rangle)^{1/2} , \tag{4.83}$$

where the inverse time is

$$\langle\tau\rangle^{-1} = \frac{(8\pi)^2 2^{1/2}(1-m)^2 n_e \hbar^2}{\Gamma^2(1/4) N_{fl} m_e^{3/2} \varepsilon_0^{1/2}} . \tag{4.84}$$

Let us estimate the value $\langle\tau\rangle$. Taking, for example, $m = 0.5$, $N_F = 10^3$, $\hbar = 10^{-27}$ erg s, $n_e = 10^{16}$ cm^{-1}, $m_e = 10^{-27}$ g and $\varepsilon_0 = 10^{-20}$ erg, we find

$$\langle\tau\rangle^{-1} = 5 \times 10^8 \, \text{s}^{-1} .$$

Thus we see that from zero up to these frequencies the formula (4.68) in accordance with (4.83) qualitatively correctly describes the parametrical dependence of the breakdown field on the frequency of the external alternating field E. The comparison with the experiments [79–81] shows quite satisfactory agreement, as illustrated in Fig. 4.5 (solid line).

4.4 Theory of Thermal Breakdown in Porous Dielectrics

The question of the thermoelectronic emission currentcaused by cathode temperature increase was first considered in the work by Richardson [82] which was written long ago and has become a classic. Since then the theoretical dependence of the current density j on the temperature obtained in that work has been named after the author. The law of variation of $j(T)$ characterizes the electric discharge in the gas phase and includes (implicitly) two cases. The first case corresponds to the condition that the length of the electron free path is of the same order or more than the distance between the electrodes, $d_{\text{electrode}}$. It is obvious that such a situation corresponds to the approximation of a highly rarefied gas in which collisions between electrons and atoms are insignificant. As to the second case, it is defined by the opposite inequality, namely $l_e \ll d_{\text{electrode}}$. It is clear that in this case it becomes significant to take into account electron scattering by the gas-phase atoms, and therefore some quite definite gas conductivity σ_g may be introduced into consideration.

The formula of Richardson has the form:

$$j = AT^{1/2} \exp\left(-\frac{\varepsilon_a}{T}\right) , \tag{4.85}$$

where the constant factor A depends on the relation $l_e/d_{\text{electrode}}$ and ε_a is the activation energy necessary for an electron to escape from the cathode surface. As in the previous section we shall use the energy system of units whereby Boltzmann's constant is held equal to one. It should be underlined that the factor A, if the condition $l_e \ll d_{\text{electrode}}$ holds, may depend on the temperature. Indeed, as the coefficient A is proportional to the conductivity σ_g, and σ_g, in turn, is proportional to the path length l_e, then $A \Rightarrow l_e \Rightarrow v_T \tau_e$. In the case of elastic collisions of electrons with atoms and molecules in the gaseous phase the probability of scattering in unit time τ_e^{-1} is proportional to $T^{1/2}$ (see, for example, the review [83]), and since the average thermal speed of electrons v_T is of the order of then (as $l_e \gg d_{\text{electrode}}$) $A = \text{const}$. If the molecular gas and electron energy density are very high, scattering with the change of the molecular internal structure may appear significant, which is connected with its transition to a new quantum state. Here the dependence of the probability τ_e^{-1} does not comply with the law $T^{1/2}$ but has a more complicated functional dependence. Without dwelling upon this point in detail we shall only note that the respective probability may be estimated by the formula $W = 1/\tau \Rightarrow T^{1/2} \exp\{-\Delta\varepsilon/T\}$, where $\Delta\varepsilon$ is the characteristic change of molecular energy accompanied by the transition either by rotational or by oscillation degrees of freedom into a new quantum state. If we assume the possibility of ionization of gas atoms by electron impact, then collisions of secondary (knocked out) electrons with primary ones will lead to a dependence of a different kind (see [84]: $W = \mu T^2$, where μ is a constant coefficient). Our reasoning, therefore, shows that in reality in most cases Richardson's formula will correctly describe both the qualitative and quantitative aspects of the thermal breakdown; however, there may be exclusions which we should not forget.

The essence of the problem solved below is the following. Suppose a two-dimensional porous dielectric is put into a weak electric field E (Fig. 4.6) and the environment temperature begins to rise. It is clear that, due to the inhomogeneity of the dielectric surface (this was much spoken about in the previous section), in some local area of its surface the average kinetic energy

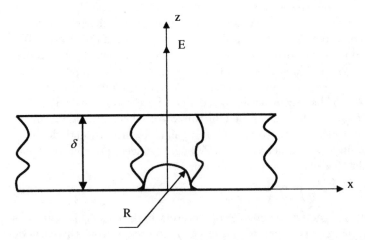

Fig. 4.6. Zone of initiation of thermal breakdown with the dimension of inhomogeneity R

of an electron will be sufficient for it to get into the dielectric volume, having overcome the potential barrier ΔV connected with the surface forces. The main direction of electron motion is naturally along the electric field E. Here, since in the path of the electron stream there necessarily will be several defects, various inhomogeneities (of the charged complex type), phonons and the like, particles scattering and losing partially their energy, will not reach the opposite boundary (the plane B–B in Fig. 4.1), and hence there will be no breakdown. In Sect. 4.2. it was shown that in the region of strong electric fields the critical field of electric breakdown is defined by the mechanism of electron scattering by fluctuations of the atomic dipole moments and (or) by the charged complexes. In the case where the experimental field E appears to be less than the critical value E^* (E^{**}), the current is absent. It is relevant to underline that those calculations which will be performed now, ultimately pursue a somewhat different purpose compared to that considered in Sect. 4.2. Namely, it is to find out the physical aspects of thermal breakdown and to determine the theoretical dependence of the breakdown current on the temperature in a porous dielectric. The mathematical method which will be basic for the solution of the stated problem is to apply the quasi-classic equation of Boltzmann to the Fermi function of the electron distribution in momentum [25], [85]. Following here the same rule as in the description of the electric breakdown field, we shall begin with the case of a homogeneous solid dielectric.

Suppose that as the environment temperature increases the phonons in the dielectric instantaneously "heat up" and become a thermostat. The latter condition imposes certain requirement on their relaxation time τ_{ph}: τ_{ph} must be much less than the electron time τ_e. And it means that the phonon distribution function is the equilibrium one:

$$\langle n_k \rangle = \left\{ \exp \left(\frac{\hbar \omega_{pk}}{T} \right) - 1 \right\}, \tag{4.86}$$

where $\omega_{pk} = c_s k$ is the phonon frequency, c_s is the sound velocity in the dielectric, k is the wave vector (we take the structure to be isotropic, whereby the sound velocity c_s is unique, i.e. $c_s^l = c_s^t$. For an estimate it is correct).

Suppose the current begins to appear at some temperature T_{cr} (its evaluation will be given below). To calculate this current let us write down the following kinetic equation:

$$\frac{\partial f}{\partial t} + v_z \frac{\partial f}{\partial z} + eE \frac{\partial f}{\partial p_z} = L_{eph}\{f, \langle n_k \rangle\}, \tag{4.87}$$

where $f(p_z, z, t)$ is the electron distribution function and $L_{eph}\{f, \langle n_k \rangle\}$ is the collision integral (see below).

Since the current density in the one-dimensional case is

$$j = \frac{eV}{(2\pi\hbar)^3 m_e} \int p_z f(p_z, z, t) \mathrm{d}^3 p, \tag{4.88}$$

the problem, therefore, is confined only to finding the electron distribution function f.

Let us find the solution of Eq. (4.87) in the simplest, but very important, stationary case. For it the following inequalities should be valid:

$$\delta t \gg \frac{\langle \delta p_z \rangle}{eE} \gg \frac{\langle v_z \rangle}{l_e} . \tag{4.89}$$

This approximation allows us to leave only the addend proportional to the electric field E. However, before we rewrite (4.87) in the approximation (4.89) we shall the collision integral in the detailed form and shall to simplify it. While the reader may find the details in the review [54], we have

$$
\begin{aligned}
L\{f, \langle n_k \rangle\} = \left(\frac{2\pi}{\hbar} \right) \sum_{p',k} |\psi(\boldsymbol{k}, \boldsymbol{p}, \boldsymbol{p}')|^2 \\
\times \{ [f_{p'}(1 - f_p)(1 + \langle n_k \rangle) - f_p(1 - f_{p'})\langle n_k \rangle] \} \\
\times \Delta(\boldsymbol{p} - \boldsymbol{p}' + \hbar\boldsymbol{k}) \delta(\varepsilon_p - \varepsilon_{p'} + \hbar\omega) ,
\end{aligned}
\tag{4.90}
$$

where the pick part (often called the scattering amplitude) is

$$\psi(\boldsymbol{k}, \boldsymbol{p}, \boldsymbol{p}') = ig \left(\frac{\hbar k}{2Mc_s} \right)^{1/2} , \tag{4.91}$$

g is the electron–phonon coupling constant [86], mass $M = \rho v_0$, ρ is the dielectric density, v_0 is the elementary cell volume and $\varepsilon_p = p^2/2m_e$ is the law of electron dispersion.

To solve the problem by the method of successive approximations we assume

$$L_{\text{eph}}\{f, \langle n_k \rangle\} = 0 .$$

Hence it follows that the equilibrium electron distribution function is

$$f = \langle f \rangle = \frac{B^*}{e^{(\varepsilon - \mu)/T} + 1} , \tag{4.92}$$

where μ is the chemical potential and the constant B^* is determined from the normalization condition

$$\left[\frac{V_{br}}{(2\pi\hbar)^3} \right] \int \langle f_p \rangle \mathrm{d}^3 p = 1 . \tag{4.93}$$

Remember that $V_{br} = \pi r_{br}^2 \delta$.

It should be stressed that the solution (4.92) and normalization (4.93) will be possible only if the collision integral $L_{\text{eph}}\{f, \langle n_k \rangle\}$ is proportional to f. It easy to verify that at high temperatures when $T \gg \langle \Theta_D \rangle$, where $\langle \theta_D \rangle$ is some characteristic temperature unit (prototype of the Debye temperature in crystals), the phonon distribution function is $\langle n_k \rangle = T/\hbar\omega_{pk}$ and then, indeed, the relation (4.90) is simplified and becomes linearly dependent on f. That is,

$$L_{\text{eph}}\{f\} = \nu \sum (f_{p'} - f_p) \delta(\varepsilon_p - \varepsilon_{p'}) , \tag{4.94}$$

where

$$\nu = \frac{\pi g^2 T}{M c_s^2 \hbar N_{br}} , \tag{4.95}$$

where N_{br} is the entire number of atoms in the breakdown area.

In the framework of the stated problem we shall look for the solution of the following equation:

$$eE \frac{\partial f}{\partial p_z} = L_{\text{eph}} \{ f, \langle n_k \rangle \} , \tag{4.96}$$

where the functional $L\{f\}$ is given by (4.94).

Let us find the solution of (4.96) in the simplest case, the so-called τ-approximation. For this purpose let us take

$$f = \langle f_p \rangle + \delta f , \tag{4.97}$$

or according to (4.96) we have

$$\delta f = -\frac{eE \tau_p \partial \langle f_p \rangle}{\partial p_z} , \tag{4.98}$$

where

$$\tau_p^{-1} = -\left. \frac{\delta L_{\text{eph}}}{\delta f} \right|_{f - \langle f \rangle} = \frac{2 \nu m_e p}{\pi^2 \hbar^3 N_{br}} . \tag{4.99}$$

Therefore, since

$$j = \left[\frac{e}{(2\pi\hbar)^3} m_e \right] \int p_z \delta f_p \mathrm{d}^3 p ,$$

substituting here (4.98) with the normalization (4.93) and also the relation (4.99) for the relaxation time τ_p we find

$$j = \left(\frac{B^* e^2 E}{m_e^3 T v_0 \nu} \right) \int_{-\infty}^{\infty} p_z^2 \mathrm{d} p_z \int_0^{\infty}$$

$$\times \frac{p \mathrm{d} p}{(p^2 + p_z^2)^{1/2} [\exp(p^2 + p_z^2)/2 m_e T + 1]} , \tag{4.100}$$

where $B^* = (2\pi)^2 \hbar^3 / \pi^{1/2} v_0 (2 m_e T)^{3/2}$ and $p^2 = p_x^2 + p_y^2$.

Consequently, taking account of the probability of the above-barrier transition into the dielectric volume $W_T = \exp(-\Delta V_0 / T)$, we finally find that the current is

$$j = \Phi T^{-3/2} \exp \left(-\frac{\Delta V_0}{T} \right) , \tag{4.101}$$

where the constant is

$$\Phi = \frac{\hbar^4 M c_s^2 e^2 E}{\delta^2 r_{br}^2 g^2 (2 m_e)^{5/2}} . \tag{4.102}$$

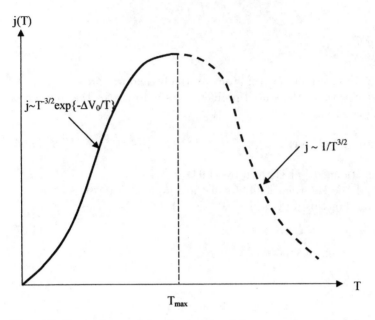

Fig. 4.7. Dependence of the thermal breakdown current on the temperature T

The dependence $j(T)$ is schematically illustrated in Fig. 4.7. Thus, indeed, as it follows from (4.101), in solid dielectrics the dependence of the thermal breakdown current differs greatly from the current of thermal-electron emission in the gaseous phase, which is described by the Richardson formula (4.85).

Let us now consider a somewhat more complicated situation. Suppose thermal breakdown occurs when a dielectric is put into a strong electric field, but not such as to cause electric breakdown. We shall find the current in this case. Note that here we cannot look for a solution of the kinetic equation in the τ-approximation, but we must find the *precise* solution of the following equation [86]:

$$eE\frac{\partial f}{\partial p_z} = \nu \sum (f_{p'} - f_p)[\delta(\varepsilon_{p'} - \varepsilon_p + \hbar\omega_{pk})\Delta(p' - p + \hbar k)$$
$$+ \Delta(p' - p' - \hbar k)\delta(\varepsilon_{p'} - \varepsilon_p - \hbar\omega_{pk})].\qquad (4.103)$$

In order to solve the given equation suppose that along the z axis the scattering by phonons is elastic, and in the x–y plane it is quasi-elastic. Mathematically this means

$$p_z = p'_z\,,\qquad (4.104)$$

$$p, p' \gg \hbar k\,,\qquad (4.105)$$

where $k = (k_x^2 + k_y^2)^{1/2}$.

We assume then that

$$f(\boldsymbol{p}_1) = f(p_z, p') = f(p_z, p + \hbar k).$$

Expanding(4.103) in terms of small values of k we obtain the following differential equation:

$$\frac{\partial f}{\partial z} + \left(\frac{b}{x^2}\right)\frac{\partial f}{\partial x} = 0, \tag{4.106}$$

where $x = p/m_e c_s$, $z = p_z/m_e c_s$ and

$$b = \frac{\nu m_e r_{br}^2 q_{max}^3 \hbar N_{br}}{12\pi (m_e c_s)^2 eE}. \tag{4.107}$$

q_{max} is the maximum phonon momentum (roughly we may take $q_{max} = \pi/\langle a \rangle$, where $\langle a \rangle$ is average interatomic distance).

Now we shall expand the function $f(x, z)$ into a Fourier integral over the variable z:

$$f = \frac{\int f_u e^{iu\tau} du}{2\pi}. \tag{4.108}$$

Substituting this expansion into (4.107), solving it and substituting, in turn, the obtained solution into (4.108), as the result of simple calculations we shall find that the sought-for electron distribution function is

$$f(p, p_z) = 2\pi D m_e c_s \delta\left[\frac{p_z - p^3}{3b(m_e c_s)^2}\right]. \tag{4.109}$$

From the normalization condition

$$\left[\frac{V_b r}{(2\pi\hbar)^3}\right]\int\limits_0^\infty\int\limits_{-\infty}^\infty f(p, p_z)p\,dp\,dp_z = 1,$$

we get

$$D = \frac{2\hbar}{m_e c_s v_0 p_{max}^2}, \tag{4.110}$$

where p_{max} is the maximum value of the electron momentum in the x–y plane.

As to the region of variation of the cross-component of momentum, it lies in the range

$$-\frac{p_{max}^3}{3b(m_e c_s)^2} \le p_z \le \frac{p_{max}^3}{3b(m_e c_s)^2}. \tag{4.111}$$

Using the obtained equations we can calculate the current of interest:

$$j = \left[\frac{eV_b r}{(2\pi\hbar)^3 m_e}\right]\int\limits_{-\infty}^\infty p_z\,dp_z\int\limits_0^\infty p f(p, p_z)\,dp.$$

After simple calculations taking account of (4.95) and (4.110) and the probability of the above-barrier tunneling W_a we finally get

$$j = \frac{3.2e^2 M c_s^2 \hbar^3 E W_a}{m_e^2 r_{br}^2 g^2 T \delta}.$$
(4.112)

As it follows from the found dependence, the qualitative behavior of the
current $j(T)$ in the region of relatively large electric fields (but less than the
field of electric breakdown) is similar to that for small E. Their difference,
which is rather significant, is only in the pre-exponential factor. Indeed, from
the comparison between (4.101) and (4.112) it follows that for weak fields
the dependence of the pre-exponential factor on the dielectric thickness is
$1/\delta^2$, and for strong fields – which the τ-approximation does not "work" –
the dependence is more smooth and is proportional to $1/\delta$.

So, the theory of the thermal breakdown of a homogeneous solid dielectric
presented above may be regarded as completed, but before we pass over
to the description of thermal breakdown in porous media we would like to
pay attention to one circumstance. This is that a number of assumptions
which simplified the mathematical transformations allowed us to bring all the
calculations to a logical completion. Obviously it was an enforced step with
the view sooner to obtain qualitative analytical equations for the temperature
dependence of the breakdown current. These assumptions enabled us not only
to calculate the current, but also to understand the physics of the process
and the character of the influence on the breakdown current of a number
of macroscopic and microscopic parameters of the system under study. The
main conclusions may be formulated in two statements:

1. The theory of thermal breakdown of homogeneous solid dielectrics has
   been considered and the breakdown current calculated
   (a) in the region of low electric fields
   (b) in strong fields less than the breakdown field.
2. We have shown the principle difference of the pre-exponential factors in
   the temperature dependence of the current in both cases and the difference between the breakdown current in dielectrics and the current of
   thermal-electron emission (cf. the Richardson formula (4.85)).

Now we shall dwell upon the current in porous structures. It is obvious
that in terms of the above material the formula for j_T may be written at once
without any intermediate transformations. Indeed, this current according to
the empirical rule given in the Sect. 4.1 (formula (4.9)) must be determined
either by the geometrical average of formulae (4.85) and (4.101) (or (4.112)),
or by the sum of these formulae multiplied respectively by m (4.85) and the
other – (4.110) (or (4.112)) – by $(1 - m)$. Thus,

$$j_T = (AC)^{1/2}[m(1-m)]^{1/2}T^{-1/2} \exp\left\{-\frac{(\Delta v_0 + \varepsilon)}{2T}\right\},$$
(4.113)

where the constant C is given by the formula (4.102).

In the case of relatively strong fields, according to (4.112), we have

$$j_T = B^{**}[m(1-m)]^{1/2}T^{-1/2}\exp\left\{-\frac{(\Delta V_0 + \varepsilon)}{2T}\right\},\qquad(4.114)$$

where $B^{**} = (ACC^*)^{1/2}$,

$$C^* = \frac{3.2e^2 M c_s^2 \hbar^3 E}{\pi m_e^2 r_{br}^2 g^2 \delta}.$$

4.5 Dependence of the Ignition Time on the Surface Temperature in Porous Dielectrics

At a first glance it may seem that the problem in question is not connected with the thermal or electric strength characteristics of porous structures. However, it is a superficial and wrong point of view.

It is well known that in order to get sound knowledge of the physics of this or that mechanism of non-equilibrium and, moreover, from the viewpoint of structure modeling it is necessary to have an idea about the various characteristics of such substances, including (and this is not an exclusion) the ignition time delay in a porous structure. Measuring the time delay we can plot (for various temperatures of surface incandescence) the corresponding curve. Then, having found a realistic theoretical interpretation of the curve complying with common sense (taking into account the functional dependence of the rest parameters interpreted earlier), we shall be able to approach gradually the most perfect model structure.

To begin with, let us experimentally study the dependence of time delay τ_d on the surface incandescence temperature T_i. The installation scheme for the τ_d measurement is shown in Fig. 4.8. As we can see from the figure, the scheme is rather simple and therefore reliable. We took six materials under test. For each material eleven measurements were made at the same fixed surface temperature.

Then the temperature T_i was changed by 30° and the test was repeated. The results of these measurements are given in the Table 4.1 and illustrated in Fig. 4.9. Processing of the numerical values shows that all of them are satisfactorily described by the power dependence in the form:

$$\tau = \frac{G}{T_i^\nu},\qquad(4.115)$$

where G is a constant and the power index varies in the range 2–8 ($2 \ll \nu \ll 8$).

Here it is very important to underline that the given empirical dependence does not fall into the framework of the traditional Arrhenius's mechanism of delay [87–90], according to which

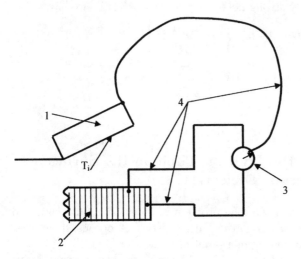

Fig. 4.8. Scheme of installation for measurement of time delays of sample ignition. 1 is incandescence surface; 2 is sample being tested; 3 is stopwatch; 4 contacts

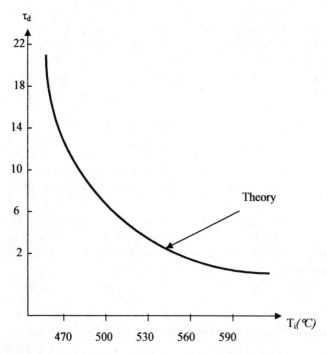

Fig. 4.9. Dependence of the time delay on the incandescence surface temperature T_i Solid line is theory, dotted line is experiment.

Table 4.1. Average time delay of various porous structures τ_d (s)

N	Type of cellulose	470	500	530	560	590
1	Cellulose for capacitor paper	11.2	8.0	6.05	5.1	4.1
2	E-2 MASHN	12.0	8.3	6.9	5.2	2.8
3	Sulfate bleached coniferous (Ust-Ilimsk)	18.1	15.23	12.3	10.8	10.1
4	PG-92 (Sweden)	10.3	7.0	5.34	4.7	2.0
5	CON-N193285 (Finland)	19.8	15.1	12.6	8.3	7.7
6	Cardicell-supercord (Sweden)	16.3	10.0	8.0	6.1	5.2

$$\tau_d = Q \exp\left(\frac{\varepsilon_a}{T_i}\right), \tag{4.116}$$

where ε_a is the activation energy connected with the atomic transition (by the energetic scale) into the zone of active chemical reactions and Q is some characteristic time.

From the comparison between these two formulae we see their qualitative difference. It should be noted however, that it will manifest itself only when the power index in (4.115) tends to the left-hand side of the inequality $2 \ll v \ll 8$. If v is close to 8, the difference from the exponential function is negligible. Thus, the task is confined to the attempt to find the interpretation of the non-trivial dependence (4.115) beyond the activation mechanism.

The difficulty of the problem is enhanced by the fact that the cellulose (or paper) type of structure burns not only outside – due to direct contact with air – but also inside – due to the internal air (gas phase) medium which saturates the pores. That is why burning is accompanied by two thermal streams: one stream $\boldsymbol{q}_e$ moves into the structure and the other one – $\boldsymbol{q}_i$ moves in the opposite direction. If the streams are similar in magnitude and opposite in direction, there occurs flame extinction in the given local area. If they are different, the intensity of burning decreases a little. Viewing all the surface of burning (e.g. by means of TV systems) we may see very strong non-uniformity of flame propagation in such disordered porous media, which is the case in practice.

To solve the submitted problem we shall choose a working hypothesis (similar to that in [91,92]) which will allow us to explain the empirical formula (4.115).

The first necessary assumption which will considerably facilitate the solution of the problem is the application of the f-points model. The second one implies that all the f-points are similar. Note that the second requirement is not compulsory, because we could introduce some function of the fibril distribution in dimensions and thus account for the non-identity of the "fibril sites". Nevertheless, we shall think that both assumptions are valid.

Let us imagine that we have isolated one surface fibril surrounded by an air layer, and assume that due to the radiant heat transfer the air medium heated instantaneously (the characteristic time τ_{rad} is much less than τ_d), and hence each fibril is put into a thermostat with the temperature T_i. The next hierarchic step is the decomposition of the fibril element. Adhering to the results of the paper [91] we shall consider that before the chemical decomposition of a fibril begins, the latter should be *prepared* to the beginning of the chemical reaction. Let us consider this point in detail. We shall postulate the following rule: if the energy level ε of chemically active structure atoms has not changed, the chemical reaction does not begin. For an atom to be prepared to the beginning of chemical transformation it needs to store some additional energy and to "jump" to the level $\varepsilon' > \varepsilon$. Active elements get the corresponding additional energy from phonons – sound vibrations of the f-point solid matrix.

Since the ignition temperature of, say, cellulose is about $200°C$, as recalculated for phonons, it means roughly that the energy difference $\varepsilon' - \varepsilon$ must be compensated by participation of two phonons in the reaction. Since it is phonons which carry the information from the fibril surface (where the temperature instantaneously became T_i) into the volume, they "inform" active atoms about the necessity to get ready for the reaction. Now it is easy to understand that it is the time of the active atom transition from the state with energy ε to the state with energy ε' which will be the time of preparation for chemical reaction and the beginning of the fibril structure decomposition. In order to present the above information in mathematical terms we shall use the approach described in the review [54] and introduce the distribution function of active atoms $\varphi(\varepsilon)$ over energy. The derivation of the kinetic equation for the distribution function $\varphi(\varepsilon)$ lies beyond the scope of the present monograph, and therefore the reader is referred to the above-mentioned review [54] for further details. As to the equation itself, it has the form:

$$\frac{\partial \varphi}{\partial t} = L\{\varphi(\varepsilon)\}, \tag{4.117}$$

where the integral of collisions may be written in the following way:

$$L\{\varphi\} = \left(\frac{2\pi}{\hbar}\right) \sum_{k,k',\varepsilon'} |A_k|^2 \{\varphi(\varepsilon')(1 - \varphi(\varepsilon))\langle n_k \rangle (1 + \langle n_{k'} \rangle)$$

$$-\varphi(\varepsilon)(1 - \varphi(\varepsilon'))\langle n_{k'} \rangle (1 + \langle n_k \rangle)\} \delta[(\omega_k - \omega_{k'})\hbar + \varepsilon' - \varepsilon], \tag{4.118}$$

where the peak part is

$$A_k = \frac{\Delta\varepsilon\hbar(ek)(ek')}{2\rho_{fl} v_{fl} c_{sfl} (kk')^{1/2}}, \tag{4.119}$$

$\Delta\varepsilon$ is the characteristic coupling energy of phonons with active elements of the fibril matrix, c_{sfl} is the sound velocity in a fibril, $v_{fl} = \pi r_{fl}^2 l_{fl}$ is a fibril volume and e is the phonon polarization.

The equilibrium Bose function of the phonon distribution is

$$\langle n_k \rangle = \left\{ \exp\left(\frac{\hbar\omega_k}{T_i}\right) - 1 \right\}^{-1}. \tag{4.120}$$

where $\omega_k = c_{sfl} k$.

As we deal with high temperatures, the distribution function at $T_i \gg \hbar \omega_\kappa$ has the form

$$\langle n_k \rangle = \frac{T_i}{\hbar \omega_k} \, . \tag{4.121}$$

Substituting this relation into the collision integral $L\{\varphi(\varepsilon)\}$ and using the explicit form of the scattering amplitude (4.119) we get

$$L\{\varphi(\varepsilon)\} = \left(\frac{\pi T_i^2 \Delta \varepsilon^2}{8 \rho_{fl}^2 v_{fl}^2 c_{sfl}^4} \right) \sum_{k,k',\varepsilon'} \left[\frac{(\boldsymbol{ek})^2 (\boldsymbol{ek'})^2}{kk'} \right] [\varphi(\varepsilon') - \varphi(\varepsilon)]$$

$$\times \delta \left[(\omega_k - \omega_{k'})\hbar + \varepsilon' - \varepsilon \right] , \tag{4.122}$$

Comparing the obtained collision integral with (4.117) we conclude that the sought-for time delay must be determined by the formula

$$\tau_d^{-1} = \left(\frac{\pi T_i^2 \Delta \varepsilon^2}{8 \rho_{fl}^2 v_{fl}^2 c_{sfl}^4} \right) \sum_{k,k',\varepsilon'} \left[\frac{(\boldsymbol{ek})^2 (\boldsymbol{ek'})^2}{kk'} \right]$$

$$\times \delta \left[(\omega_k - \omega_{k'})\hbar + \varepsilon' - \varepsilon \right] . \tag{4.123}$$

In order to perform the summation with respect to k, k' and ε' we use the rules

$$\sum_k (\ldots) = v_{fl} \int (\ldots) \mathrm{d}^3 k \, ,$$

$$\sum_{\varepsilon'} (\ldots) = \int \mathrm{d}\varepsilon' n(\varepsilon')(\ldots) \, ,$$

where the density of states of the active atomic system is

$$n(\varepsilon') = \mathrm{d}N(\varepsilon')\mathrm{d}\varepsilon' \, .$$

Holding that $n(\varepsilon') = \Delta E^{-1}$, where ΔE is the characteristic energy spread in an active atomic system, after simple transformations, using the properties of the δ-function we find

$$\tau_d^{-1} = \frac{\pi T_i^2 \Delta \varepsilon^2 (N_{afl})^2}{24 \hbar \rho_{fl}^2 v_{fl}^2 c_{sfl}^4 \Delta E} \, , \tag{4.124}$$

where N_{afl} is the entire number of atoms in a fibril.

If now we introduce the only adjustment parameter

$$\varepsilon_0^* = \frac{\Delta \varepsilon^2}{\Delta E} \, , \tag{4.125}$$

we may rewrite the functional dependence of the time delay in the form

$$\tau_d = \frac{24 \rho_{fl}^2 v_{fl}^2 c_{sfl}^4 \hbar}{\pi T_i^2 \varepsilon_0^* N_{afl}^2} \, . \tag{4.126}$$

Finally, taking into consideration that we have an ensemble of fibrils whose number is $N_{fl} = (1 - m)V_{\min}/v_{fl}$, where $V_{\min}$ is the minimum volume of the local ignition area where the burning begins we ultimately obtain

$$\tau_d = \frac{G}{T_i^2}, \qquad (4.127)$$

where the constant is

$$G = 24\rho_{fl}^2 v_{fl} V_{\min}(1 - m)\frac{c_{sfl}^4 \hbar}{\pi \varepsilon_0^* N_{afl}^3}. \qquad (4.128)$$

Note, that each individual structure has its own $V_{\min}$.

Thus, from the obtained dependence it follows that the time delay τ_d decreases in those samples where

(a) the pore volumes are larger;
(b) the sound velocities in a fibril are lower;
(c) the fibril volumes are smaller.

And the tendency to increase is observed in such samples where

(a) the fibril structure density ρ_F is higher;
(b) the primer energies ε_0^* are lower.

Comparison of the obtained dependence with the experimental points shows that they are in quite satisfactory agreement (see Fig. 4.9, where the solid line corresponds to the theoretical dependence (4.127)). For the complete quantitative agreement of theory with experiment the magnitude of the primer energetic parameter ε_0^* should be chosen correctly.

4.6 The Influence of the Viscosity of the Saturating Liquid on the Electric Strength of Porous Dielectrics

We have already discussed in Sect. 4.3 how the viscosity of the impregnating liquid influences the electric strength of a porous dielectric.

However, in the above-mentioned section we only briefly touched upon this question and did not specially draw the reader's attention to it.

Due to the importance for practical use of, for example, capacitor paper, we shall now thoroughly investigate the question of how the viscosity of the liquid can influence its electrical strength.

As we know, on exposure of the gas and liquid of in a solid dielectric to an electric field, their "resistance" to a breakdown increases in the same sequence, i.e. $E_{br}^{(g)} < E_{br}^{(l)} < E_{br}^{(sol)}$.

The conclusion is obvious: if we have a porous dielectric and want to increase its electric strength, we should impregnate it with some liquid compound (e.g. camphor oil) whose density is higher, naturally, than that of the gas.

Further, as the density is

$$\rho = (1 - m)\rho_{fl} + m\rho_g \,,$$

hence

$$\rho_g < \rho < \rho_{fl} \,. \tag{4.129}$$

If we now impregnate paper with liquid of density η_l, the inequality (4.129) will be modified and will give the condition:

$$\rho_g < \rho_l < \rho^{br} < \rho_{fl} \,. \tag{4.130}$$

Therefore, the complete density will become equal to

$$\rho^{br} = (1 - m)\rho_{fl} + m\rho_l \,. \tag{4.131}$$

On the other hand, from experience we know (see Chap. 3) that

$$E^* \Rightarrow \rho^\alpha \,,$$

where $\alpha \geq 1$, and according to (4.131) we see that the electric strength of the impregnated paper increased. Thus we have assured ourselves that the viscosity increases the breakdown field of a porous substance. From (4.64) we have

$$E_{br} = \left(\frac{6\eta_e l_{fl}}{\rho_{fl} r_{br}^2} + \frac{m\eta_g}{\rho_e \delta} \right)^2 \frac{\rho_e r_{br}^2}{e} \,,$$

and consequently

$$E_{br} = A_1 + A_2\eta_l + A_3\eta_l^2 \,, \tag{4.132}$$

where the constant parameters $A_{1,2,3}$ are given by the previous relation.

The dependence plot of E_{br} against the liquid viscosity is shown in Fig. 4.10. From this figure we can see that in the region of low viscosity the breakdown field may be described by a linear dependence from $\eta_l \equiv \eta$ increasing, it changes into a stronger square-law dependence.

Returning again to the formula (4.64) we conclude, based on the relations $\eta_g = \rho_g v_g$, $\rho_g = \rho/m - (1 - m)\rho_{fl}/m$ and on the condition of constant fibril density, that the breakdown field is

$$E_{br} = B_1 + B_2\rho + B_3\rho^2 \,, \tag{4.133}$$

where the constant coefficients $B_{1,2,3}$ are determined from (4.64).

Relation (4.133) is in rather good agreement with experiments studying the breakdown field dependence on the structure density (see, for example, the monograph [93]; Fig. 4.11 convincingly illustrates this (see also Sect. 3.5).

Thus, we have assured ourselves that in order to get capacitor paper having higher electric strength its structure should be impregnated with the *most* viscous liquid. And though this circumstance was found in practice, there was, for some reason, no analytical foundation. We have bridged this gap by a detailed presentation of the physical pattern of the phenomenon.

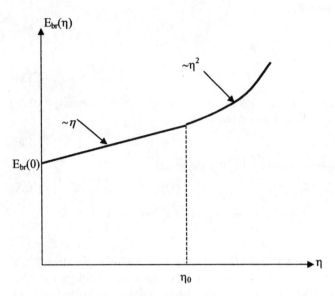

Fig. 4.10. Plot of the electric breakdown field versus the liquid viscosity η

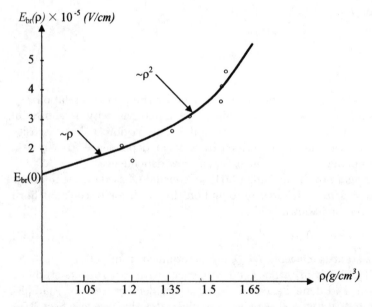

Fig. 4.11. Breakdown field as a function of the sample density. The solid curve is theory; the circles are experiment.

Let us now sum up the sections presented above:

1. Formulae (4.35), (4.48) and (4.67) describing the functional dependence of the electric breakdown field of porous substances on various structure parameters allow us to see the breakdown process from different points of view, each being able to interpret vast experimental data.
2. The dependence has been found of the break force as a function of the structure density and the treatment of many experiments has been given.
3. The dependence of the breakdown field on the viscosity of the saturating liquid η and the structure density ρ has been described.
4. Some hypothetical methods to improve the electric strength and the thermal characteristics of porous structures have been suggested.
5. A non-trivial interpretation of the cellulose ignition time delay as a function of the incandescent surface temperature has been given.
6. The method of solution of the one-dimensional Boltzmann kinetic equation has been demonstrated and methods have been suggested to calculate the coefficient of electric conductivity in a porous substance from the frequency of the external alternating electric current.

4.7 Some Aspects of Relaxation Theory in Porous Dielectrics

Before we start to investigate the coefficient of thermal conductivity of fiber–fibril structures as one of the most complicated classes of inhomogeneous and greatly non-ordered substances, it is necessary to familiarize the reader with the physical fundamentals and calculation methods, and, besides, with estimation of the relaxation times on the basis of a simpler (than that presented in Chap. 5) theoretical analysis. We shall deal with the application of the Boltzmann kinetic equation (which we used in the calculation of the electric breakdown current of a porous structure in Sect. 4.2) for estimation of the heat-conducting properties of such media. Moreover it should be done, because from the estimation point of view the kinetic equation is the unique instrument and calculation apparatus in an overwhelming majority of cases when non-equilibrium microscopic processes are investigated. Here, however, it should be stressed that such an analysis is possible only in the quasi-classically approximation when the condition for the quasi-classical approach is satisfied, i.e. when the product of the length interval Δx an the momentum change Δk is considerably larger than one. The role of Δx for particles (or quasi-particles) scattering from each other is played by their free path length l, and the momentum change Δk in collision processes corresponds to the momentum of the scattered particle (or quasi-particle) k. Therefore, the condition of applicability of the quasi-classical approach may be written in the most customary form for us, namely $kl \gg 1$. If we denote the group velocity by u and introduce the time τ between collisions of quasi-particles (we shall deal only with them), then the stated condition may be presented in the equivalent form $ku\tau \gg 1$. On the other hand, the product of k and u is nothing but the product $\Delta k(\partial\omega/\partial k)$, where $\omega(k)$ is the quasi-particle frequency and hence the magnitude $\Delta k(\partial\omega/\partial k)$ is the quasi-particle frequency change in the scattering process on the magnitude $\Delta\omega(k)$. If $\Delta\omega$ corresponds by order of magnitude to ω, the condition of quasi-classical approximation instead of inequality $\Delta\omega(k)\tau \gg 1$ must be written as

$$\omega(k)\tau \gg 1. \tag{4.134}$$

The condition (4.134) is nothing but the condition of applicability of the Born approximation for the calculation of the relaxation time τ!! Note, by the way, that if we deal with purely quantum calculations, such an approximation is the called "Fermi golden rule" and the role of the dispersion change $\Delta\omega(k)$ moves to the energy level difference of the system under investigation. The time τ in this case will characterize level widening as a result of some interactions.

As we noted above, the description of relaxation processes even in the simplest crystal structures is connected with very awkward calculations of a purely mathematical character demanding that we take into account the symmetry properties of the crystal lattice, and, of course, maximum attention in analyzing the various microscopic mechanisms of internal energy dissipation of the class of substances under investigation. The quantity of relaxation mechanisms increases nonlinearly depending on the type of physical construction of a structure and which (of course, absolutely conditionally) may be arranged in the form of the following hierarchic series according to degree of complexity: dielectrics, magnets, metals, ferromagnetic metals, semiconductors, alloys, composites, porous structures, and finally porous composite structures. Here there is always a real probability of missing some very effective mechanisms of interaction (with the smallest relaxation time) which thus play a rather considerable role in the qualitative character of reaching of internal equilibrium state. Failure to take such mechanisms into account will radically change (to the wrong side) the whole character of establishment of thermodynamic balance in the subsystems under consideration.

It should be noted that only correct relaxation theory allows us to develop a correct theory of the absorption of electric field energy in dielectrics (see Chap. 6), which will essentially give the correct temperature dependence of the dielectric permeability $\varepsilon(\omega,T)$, where ω is the external electric field frequency.

This statement holds for any other class of the listed structures and, in particular, for magnets with respect to finding the frequency and temperature dependence of their magnet susceptibility.

As one of the most complicated classes (but much simpler the than fiber–fibril substance) the porous crystal dielectric is an ideal object for investigation. It was chosen by the author only as illustrative example and only with the purpose of demonstrating the whole analysis of relaxation mechanisms and the whole algorithm of description of setting up of internal thermodynamic equilibrium in such very inhomogeneous media.

As we have already stressed, the relaxation theory is a necessary and sufficient condition for the adequate description of the temperature dependence of the thermal conductivity coefficients of corresponding substances.

So, the pattern of relaxation is closely connected with the establishment of heat balance in a pure dielectric, which is why we shall start our consideration with it.

The traditional mechanism of the final stage of achieving equilibrium in dielectrics was and, so far is considered to be, the change of state mechanism (the so-called umklapp processes), which was first introduced into present terminology by R. Peierls in 1929. It should be noted that taking into account the change of state process we cannot ignore the interrelation of the non-equilibrium subsystems with the thermostat in which some constant temperature T_0 is maintained. The interaction of the volume phonons with the surface phonons (already was thermostat due to their connection with the external environment whose temperature is T_0) in reality plays a very important role and leads to the final equalization of all non-equilibrium parameters. It means that the quasi-equilibrium temperature of volume phonons T tends to T_0, and the chemical potential of phonons μ and the velocity of phonon entrainment V tend to zero.

Now we shall try to describe and at least qualitatively to estimate the character of establishment of internal thermal microscopic equilibrium by analysis and isolation of the main dissipation mechanisms of energy and momentum in the following subsystems: (a) longitudinal phonons (l), (b) transverse phonons (t), (c) optical phonons (O) and (d) a thermostat maintaining temperature $T_0 = $ const.

Suppose, that the phonon system is disturbed from the equilibrium state and after it the dielectric is immediately put into the thermostat. In order to describe microscopically how the equilibrium distribution in subsystems (a), (b) and (c) is established, we shall adhere to the following principle. We shall believe that the smallest relaxation time corresponds to a three-particle scattering process in the system of transverse phonons (t), which, in principle, is quite clear, because the density of states in the system of transverse phonons is a maximum. Let us denote this time as τ_{ttt}. The next in the hierarchic chain of times will be the time τ_{llt}, corresponding to the interaction processes $b^+_{lk} b_{lk'} b^+_{tk''}$, where the operators $b^+_{lk}(b_{lk})$ are the operators (elimination) of a longitudinal phonon (l) with wave vector k, and $b^+_{tk}(b_{tk})$ is the same operator but for transverse phonons (t). The process competing with this interaction mechanism is impurity scattering of a phonon (l) with its subsequent transformation into a transverse phonon (t). The relaxation time for this process will be denoted as τ_{implt}.

Then, the time for the establishment of the quasi-equilibrium distribution of the longitudinal phonons τ_{lll} turns out to be the largest one from the listed times, but, however, less than relaxation time of a longitudinal phonon accompanied by the process of $b_{lk}^+ b_{Ok'}^+ b_{Ok''}$-type with participation of optical phonons (index "O"). The corresponding time will be denoted as τ_{lOO}. Closing the enumerated hierarchic chain of times we should now introduce the slowest relative to the above mechanisms process of relaxation connected with the change of state processes and characterized by the time τ_{lU}. The above times may be united into one mathematical inequality of the type

$$\tau_{ttt} \ll \tau_{llt} < \tau_{\text{limplt}} < \tau_{lll} \ll \tau_{lOO} \ll \tau_{lU} . \tag{4.135}$$

Note, that the given chain of inequalities between the average times can be strictly proved only by numerical integration of the corresponding analytical expressions, which we shall give below. Let us now accept the inequalities (4.135) as a fact and by means of them find out only the qualitative pattern of all the subsystems attaining a state of thermal equilibrium.

So, the relaxation pattern must be the following. In time τ_{ttt} the transverse phonons obtain their temperature T_t and their quasi-equilibrium distribution function will be

$$\langle N_{tk} \rangle = \left\{ \exp \left[\frac{(\hbar\omega_{tk} - \hbar \mathbf{k} \mathbf{V}_t)}{T_t} \right] - 1 \right\}^{-1} , \tag{4.136}$$

where $\mathbf{V}_t$ is the average velocity of the stream entrainment of the transverse phonons, whose dispersion is $\omega_{tk} = c_t k$, where c_t is their phase velocity.

Then, due to the interaction $b_{lk}^+ b_{lk'} b_{tk''}^+$ which commutes with the operator of the number of particles of the longitudinal phonons, i.e. with $N_l = b_{lk}^+ b_{lk}$, the longitudinal phonons acquire temperature T_t, stream velocity of transverse phonons $\mathbf{V}_t$ and *not equal to zero* chemical potential μ_l, and, therefore, their quasi-equilibrium distribution function is as follows:

$$\langle N_{lk} \rangle = \left\{ \exp \left[\frac{(\hbar\omega_{lk} - \hbar \mathbf{k} \mathbf{V}_t - \mu_l)}{T_t} \right] - 1 \right\}^{-1} , \tag{4.137}$$

where the dispersion of the longitudinal phonons is $\omega_{lk} = c_l k$, and c_l is their phase velocity.

Now let us recollect that there is the thermostat of transverse (the fastest!) phonons next to (in some abstract region Ω_δ, where δ is the distance counted from a dielectric boundary inwards and equal in order of magnitude to the maximum length of the phonon free path, i.e. $\delta \approx l_{\max} = c_l \tau_{\max}$) the thermostat T whose temperature is maintained constant and equal to T_0. For phonons from the boundary region Ω_δ, the time of the three-particle interaction process is τ_{TTT}, less than or of the order of τ_{ttt} and for which the equilibrium Bose distribution has the form of a purely Planck distribution:

$$\langle N_{Tk} \rangle = \left[\exp \left(\frac{\hbar\omega_{tk}}{T_0} \right) - 1 \right]^{-1} . \tag{4.138}$$

That is why in addition to the inequality (4.135) one more parallel chain of inequalities should be written

$$\tau_{TTT} < \tau_{ttt} \ll \tau_{ttT(t)} \ll \tau_{lT(l)t} < \tau_{\text{limp}lT(t)} \ll \tau_{lU} \,. \tag{4.139}$$

Here we shall underline that the lower index of relaxation times written down in the form of signs $T(t)$ and $T(l)$ means that here the interaction is accounted with one (one sign T!) transverse ($T(t)$) or one longitudinal ($T(l)$) phonon.

If we denote the linear dimension of a sample by L, then there is the following simple relation between the times:

$$\begin{cases} \tau_{ttT(t)} = \left(\dfrac{L}{2\delta}\right)\tau_{ttt}\,, \\[2ex] \tau_{lT(l)t} = \left(\dfrac{L}{2\delta}\right)\tau_{llt}\,, \\[2ex] \tau_{\text{limp}lT(t)} = \left(\dfrac{L}{2\delta}\right)\tau_{\text{limp}lt}\,. \end{cases} \tag{4.140}$$

Now in order to describe strictly mathematically the establishment of an equilibrium distribution in the subsystems under investigation it is necessary to introduce the quasi-classic kinetic equation for the longitudinal and transverse phonons. In the general case they will have the following form. For longitudinal phonons:

$$\dot{N}_{lk} = L\{N_{lk}, N_{tk}\}\,, \tag{4.141a}$$

for transverse phonons:

$$\dot{N}_{tk} = L\{N_{tk}, N_{lk}\}\,, \tag{4.141b}$$

where the functionals $L\{N\}$ on the right-hand sides of (4.141a) are collision integrals which we already know from the above material.

Functions N_{lk} and N_{tk} imply that we include the interaction between longitudinal and transverse phonons.

As to the total time derivative which is on the left-hand side of the given equations, its detailed expression, say, for longitudinal phonons may be presented as

$$\dot{N}_{lk} = \left(\frac{\partial N_{lk}}{\partial t}\right) + \boldsymbol{v}\nabla N_{lk} + \dot{\boldsymbol{p}}\left(\frac{\partial N_{lk}}{\partial \boldsymbol{p}}\right)\,, \tag{4.142}$$

where the vector $\boldsymbol{v}$ characterizes the group velocity of a quasi-particle and the vector $\boldsymbol{p}$ characterizes its momentum.

An analogous relation holds for the distribution function of the transverse phonons.

The system of Eqs. (4.141a) and the detailed Eq. (4.142) for the total time derivative from the distribution function allows us, in turn, to obtain the system of equations describing the relaxation of the macroscopic parameters $\boldsymbol{V}_t$, μ_l and the difference $\delta T = |T_t - T_0|$, which is much smaller than T_0.

Indeed, based on the equalities (4.139), we may now write the kinematic equations in the so-called τ-approximation, which is quite sufficient for our purpose. In the linear approximation over the independent parameters $\boldsymbol{V}_t$, μ_l

and the small difference $\delta T = |T_t - T_0|$-approximation for transverse phonons
we shall find the following equation:

$$
\left(\frac{\partial\langle N_{tk}\rangle}{\partial T_t}\right)\Bigg|_{\substack{T_t = T_0 \\ \boldsymbol{V}_t = 0}} \delta \dot{T} - \left(\frac{\partial\langle N_{tk}\rangle}{\partial(\hbar\omega_{tk})}\right)\Bigg|_{\substack{T_t = T_0 \\ \boldsymbol{V}_t = 0}} \hbar\,\boldsymbol{k}\dot{\boldsymbol{V}}_t
$$

$$
= \left(\frac{\delta L_{ttT(t)}}{\delta N_{tk}}\right)\Bigg|_{N_{tk}=\langle N_{tk}\rangle} \times \Bigg\{ \delta T \left(\frac{\partial\langle N_{tk}\rangle}{\partial T_t}\right)\Bigg|_{\substack{T_t = T_0 \\ \boldsymbol{V}_t = 0}}
$$

$$
- \hbar\boldsymbol{k}\boldsymbol{V}_t\left(\frac{\partial\langle N_{tk}\rangle}{\partial(\hbar\omega_{tk})}\right)\Bigg|_{\substack{T_1 = T_0 \\ \boldsymbol{V}_t = 0}} \Bigg\} - \left(\frac{\boldsymbol{k}\boldsymbol{V}_t}{\omega_{tk}}\right)\left(\frac{1}{\tau_{lUk}}\right), \tag{4.143}
$$

where $1/\tau_{lUk}$ is probability of phonon scattering per unit time due to change
of state processes and $L_{ttT(t)}$ is the collision integral of the transverse phonon
of the thermostat from the region $\Omega_\delta(T(t))$ with two transverse phonons (t)
from the dielectric volume.

Quite similarly, using the quasi-equilibrium distribution (4.137), we find
by means of the expansion of the Bose function over the corresponding small
parameters:

$$
\left(\frac{\partial\langle N_{lk}\rangle}{\partial T_t}\right)\Bigg|_{\substack{T_t = T_0 \\ \boldsymbol{V}_t = 0 \\ \mu_l = 0}} \delta \dot{T} - \left(\frac{\partial\langle N_{lk}\rangle}{\partial(\hbar\omega_{tk})}\right)\Bigg|_{\substack{T_t = T_0 \\ \boldsymbol{V}_t = 0 \\ \mu_l = 0}} \hbar\boldsymbol{k}\,\dot{\boldsymbol{V}}_t
$$

$$
= \left(\frac{\delta l_{lT(l)t}}{\delta N_{tk}}\right)\Bigg|_{N_{lk}=\langle N_{lk}\rangle} \times \Bigg\{ \delta T \left(\frac{\partial\langle N_{lk}\rangle}{\partial T_t}\right)\Bigg|_{\substack{\boldsymbol{V}_t = 0 \\ T_t = T_0 \\ \mu_l = 0}}
$$

$$
- \hbar\boldsymbol{k}\boldsymbol{V}_t\left(\frac{\partial\langle N_{lk}\rangle}{\partial(\hbar\omega_{tk})}\right)\Bigg|_{\substack{T_t = T_0 \\ \boldsymbol{V}_t = 0 \\ \mu_l = 0}} - \left(\frac{\partial\langle N_{lk}\rangle}{\partial(\hbar\omega_{tk})}\right)\Bigg|_{\substack{T_t = T_0 \\ \boldsymbol{V}_t = 0 \\ \mu_l = 0}} \times \mu_l \Bigg\}
$$

$$
- \left(\frac{\boldsymbol{k}\boldsymbol{V}_t}{\omega_{lk}}\right)\left(\frac{1}{\tau_{lUk}}\right), \tag{4.144}
$$

where $L_{lT(l)t}$ is the integral of collision of two volume phonons with a transverse phonon of the thermostat from the known region Ω_δ.

By definition, the variational derivatives appearing in (4.143) and (4.144) enable us to introduce two relaxation times. That is:

$$\left.\frac{\delta L_{ttT(t)}}{\delta N_{tk}}\right|_{N_{tk}=\langle N_{tk}\rangle(T_t=T_0,\boldsymbol{V}_t=0)} = -\frac{1}{\tau_{ttT(t)k}}, \qquad (4.145)$$

and

$$\left.\frac{\delta L_{lT(l)t}}{\delta N_{lk}}\right|_{N_{lk}=\langle N_{lk}\rangle(T_t=T_0,\mu_l=\boldsymbol{V}_t=0)} = -\frac{1}{\tau_{lT(l)tk}}, \qquad (4.146)$$

From inequalities (4.139) we make the conclusion that the final establishment of equilibrium will be defined only by the Eq. (4.144). Really, due to independence of the parameters μ_l, δT and $\boldsymbol{V}_t$, from (4.144) we can easily obtain the following three equations:

$$\begin{cases} \dot{\delta T} = -\tau_1^{-1}\delta T\,, \\ \dot{\mu}_l = -\tau_2^{-1}\mu_l\,, \\ \dot{\boldsymbol{V}}_t = -(\tau_3^{-1}+\tau_U^{-1})\boldsymbol{V}_t\,, \end{cases} \qquad (4.147)$$

where the average relaxation times are defined by the equations

$$\frac{1}{\tau_1} = \frac{\int \tau_{lT(l)tk}^{-1}(\partial\langle N_{lk}\rangle/\partial T_t)\Big|_{T_t=T_0,\mu_l=\boldsymbol{V}_t=0}\times \mathrm{d}^3k}{\int(\partial\langle N_{lk}\rangle/\partial T_t)\big|_{T_t=T_0,\mu_l=\boldsymbol{V}_t=0}\times \mathrm{d}^3k}, \qquad (4.148)$$

$$\frac{1}{\tau_2} = \frac{\int \tau_{lT(l)tk}^{-1}(\partial\langle N_{lk}\rangle/\partial(\hbar\omega_{lk}))\Big|_{T_t=T_0,\mu_l=\boldsymbol{V}_t=0}\times \mathrm{d}^3k}{\int(\partial\langle N_{lk}\rangle/\partial(\hbar\omega_{lk}))\big|_{T_t=T_0,\mu_l=\boldsymbol{V}_t=0}\times \mathrm{d}^3k}, \qquad (4.149)$$

$$\frac{1}{\tau_3} = \frac{\int k^2\tau_{lT(l)tk}^{-1}(\partial\langle N_{lk}\rangle/\partial(\hbar\omega_{lk}))\Big|_{T_t=T_0,\mu_l=\boldsymbol{V}_t=0}\times \mathrm{d}^3k}{\int k^2(\partial\langle N_{lk}\rangle/\partial(\hbar\omega_{lk}))\big|_{T_t=T_0,\mu_l=\boldsymbol{V}_t=0}\times \mathrm{d}^3k}, \qquad (4.150)$$

It should be noted that the Eqs. (4.147) are obtained in the following way. The first equation expressing the law of phonon energy conservation was found by multiplication of both sides of the equality (4.144) by $\hbar\omega_{lk}$ and subsequent integration with respect to d^3k in the limits of the first Brillouin zone. The second equation, which expresses the law of conservation of quasi-particles, was found simply by integration of both sides of the equality (4.144) over the whole region $\boldsymbol{k}$, which is also in the limits of the first Brillouin zone.

Finally, the last equation defining the velocity relaxation of the phonon stream and expressing the law of momentum conservation of the whole quasi-particle system is obtained by multiplication of (4.144) by the vector $\boldsymbol{k}$ and then (as in two previous equations) both sides are integrated with respect to d^3k. Here, in the latter case, as it is quite obvious, addends proportional to parameters δT and μ_l automatically fall out from the equation on the velocity $\boldsymbol{V}_t$ due to angle integration.

Therefore the latter expression describes the momentum relaxation of the phonon stream in which there must be and there is average time of change of state processes τ_U.

In the most interesting for investigation (both theoretical and experimental) range of temperatures, where $T < \theta_D$, the condition $1/\tau_3 \gg 1/\tau_U$ will be valid, and hence the time of ultimate establishment of equilibrium in all the subsystems by order of magnitude will give the value $\tau_{lT(l)t} \approx (L/2\delta)\tau_{llt}$. If a sample is small, then $L \geq 2\delta$ and $\tau_{lT(l)t} > \tau_{llt}$. If a sample is massive, the dimension L considerably exceeds 2δ and the time $\tau_{lT(l)t}$ becomes much larger than τ_{llt}. Here the role of change-of-state processes increases sufficiently and for massive samples they will play the main role.

Here it would be relevant to say some words concerning the parameter δ. By definition the layer of a dielectric adjacent to the surface is a thermostat and its width must be defined by the small relaxation time of the surface (though three-dimensional!) phonons. It means that the value $\delta \approx c_l\tau_{ltt}(T_0)$ because $\tau_{ltt}(T_0)$ is τ_{max}. Numerical estimations show that the time τ_{max} is about 10^{-7}–10^{-6} s, and hence the width of the transitional layer will be about $10^6 \times 10^{-7} = 0.1$ cm (here we have taken the sound velocity c_l to be roughly equal to 10^6 cm/s). For samples with sizes from 2 to 5 mm the smallness parameter $2\delta/L$ will be, therefore, in the range from 0.4–1.0.

The given estimations and reasoning allow us to construct the qualitative pattern of relaxation in a homogeneous crystal dielectric. Its essence is the following. During the time τ_{ttt} in the system of transverse phonons its own quasi-equilibrium temperature T_t and also entrainment velocity of stream of transverse phonons $\boldsymbol{V}_t$ not equal to zero have been established. In time of order of τ_{llt} longitudinal phonons adjusted themselves to transverse ones and acquired the temperature of transverse phonons T_t, their velocity $\boldsymbol{V}_t$ and chemical potential μ_l not equal to zero. The next stage is the interaction of transverse phonons with thermostat transverse phonons from the region Ω_δ (the time has the value of order of $\tau_{ttT(t)}$), which results in their own equilibrium temperature T_0 and phonon entrainment velocity equal to zero ($\boldsymbol{V}_t = 0$). And, finally, in time $\tau_{\lim plT(t)}$ the ultimate establishment of thermal equilibrium takes place in all the subsystems: the chemical potential of longitudinal phonons tends to zero, their velocity of phonon entrainment also tends to zero, and their quasi-equilibrium temperature which is equal to the temperature of transverse phonons T_t tends to the thermostat temperature T_0. The described procedure is the qualitative theory of relaxation in a crystal dielectric. Knowing it we shall be able to pass over to more complicated structures, such as the porous dielectric. In order to create a relaxation theory for a porous structure we should take some model as a basic one and construct relaxation taking account of the porosity m.

So, having learned the fundamentals of relaxation theory in porous substances, let us somewhat complicate the problem and again, as an example, consider the question of the establishment of an equilibrium thermodynamic state in a crystal dielectric, but when in its structure there is only one single macro-inhomogeneity in the form of spherical cavity with radius R.

It is clear that for small pore dimensions, when the inequality $R \ll L$ is valid, the relaxation will remain practically the same as in a homogeneous dielectric and the presence of a pore will not greatly influence its physical properties. With the increase of R the tendency will appear to domination of pore's properties and the main non-equilibrium mechanism will be the interaction not between phonons in the main matrix, but the interaction between photons and phonons in the vicinity of the pore surface.

Suppose that as a result of the action of some external radiation source we disturb the whole internal structure of the dielectric from the equilibrium state and then instantly place it into the thermostat with temperature T_0. Then at once a reasonable question arises: how to describe the thermostating process of all the internal subsystems? We shall start from the following. Let all the subsystems be in equilibrium, and in a pore the so-called equilibrium black-body radiation is established issuing from the dielectric into the pore. Its energy according to the Stefan–Boltzmann law is $E_2 = (4V_2/c)\sigma T_0^4$, where $V_2 = 4\pi R^3/3$ is the volume of the spherical region, c is the velocity of light in vacuum, and the coefficient σ is the Stefan–Boltzmann constant, equal to $\sigma = (\pi^2/60)(k_B^4/h^3c^2) \cong 5.67 \times 10^{-5}\,(\mathrm{gs}^{-3}\mathrm{K}^4)$.

Now let the equilibrium be disturbed and the black-body radiation inside the pore became non-equilibrium. What then? In compliance with the Bouguer–Lambert law (according to which radiation intensity J from the gaseous-phase region (from the pore) will begin to move into the dielectric and there, in turn, will be lost according to the law $J_0 \mathrm{e}^{-x/h}$, where J_0 is the initial radiation at $x = 0$ (on the sphere surface) and h is the scattering depth), non-equilibrium electromagnetic (EM) radiation will begin to penetrate into the main matrix of the dielectric, and scattering by fluctuations of density will relax to the equilibrium state with energy E_2. The penetration depth h will be determined as the product of light velocity in the substance $c^* = c/(\varepsilon\mu)^{1/2}$, where ε is the dielectric permeability and μ is the magnetic permeability, and the relaxation time τ connected with the processes of interaction of the EM field quanta (photons) with the oscillations of the crystal lattice atoms, the phonons. So, formally, $h = c\tau/(\varepsilon\mu)^{1/2}$. Interaction of non-equilibrium EM radiation takes place in the region directly neighbouring the pore, whose volume may be determined as the difference $V_h = V_{h+R} - V_R = (4\pi/3)[(R + h)^3 - R^3] = (4\pi/3)h(h^2 + 3Rh + 3R^2)$. Here it should be noted that at small h (weak interaction) the contact region becomes equal to $4\pi h R^2$ $(R \gg h)$, and at relatively small radius the situation becomes very realistic when h by order of magnitude corresponds to R, and the contact region between non-equilibrium photons and phonons will be defined by the general expression V_h.

Thus we understood that h is some average length of the photon free path up to its complete annihilation in the substance. Since, in general, we

are dealing with the infrared band of EM waves, their scattering directly
by the crystal lattice atoms of the dielectric will be elastic (for the photon
wavelength λ_{phot} considerably exceeds the atom dimensions), and hence for
the relaxation pattern on the whole it is quite insignificant because it does
not lead to relaxation over energies or momentum.

Before we pass over to a qualitative description of the relaxation process
in the two-phase structure under consideration, one more thing should be
discussed. In estimating the relaxation times, of great importance is the question of what relaxation time is calculated. For example, let us consider a
phonon. The matter is that in finding the relaxation time of a phonon during
the act of its inelastic collision with a photon it is necessary in the process
of integration with respect to the virtual momentum of the photons to introduce their density of states. The density of states, as we know, is defined by
the number of particles in the given interval of energies (momentum). Due
to the fact that the phase propagation velocity of a photon is the velocity
of light (and hence its energy will be proportional to c), the density of state
for a photon turns out to be inversely proportional to c^3 and in integration
with respect to the virtual region of their (photons') wave vectors, the inverse
relaxation time will appear proportional to c^{-3} (i.e., $\tau_{\text{phon}-\text{phot}k}^{-1} \sim c^{-3}$).

When we deal with the photon relaxation time for the same scattering
process with the participation of the phonons, it is necessary to perform the
summation over the virtual wave vectors of the phonons, which is equivalent
to the introduction of their density of states, which obviously will be inversely
proportional to cube of the sound velocity. Thus, the inverse relaxation time
for a photon will be $\tau_{\text{phot}-\text{phon}k}^{-1} \sim c_s^{-3}$. From the comparison of these times
the obvious conclusion may be made that the relaxation time of the photons
turns out to be much smaller (by a factor $(c_s/c)^3$) than that of phonons for
the same process.

Having cleared up the given point we may now pass over to the qualitative
description of the process of establishing thermodynamic equilibrium in the
subsystems under consideration.

So, due to the presence of such parameters as the light velocity, we undoubtedly may state the domination (on a hierarchic timescale) of the inter-
phonon interactions over processes with photon participation. This enables
us to write the chain of inequalities between the corresponding and already
known to us relaxation times. Indeed, inequality (4.139) will include two ad-
ditional components, namely, the inverse average time $\tau_{\text{phot}-\text{phon}}^{-1} \equiv \tau_{ph\,ph-t}^{-1}$
and τ_{ph-t-t}^{-1}, (index "ph" meaning photon). Here we have underlined that
the main mechanisms are photon interactions not with longitudinal but with
transverse phonons for the reason submitted above, which is again connected
with the density of states. Summing up, we have

$$\tau_{TTT} < \tau_{ttt} \ll \tau_{ttT_t} \ll \tau_{lT_l t} < \tau_{\text{limpl}T_t} \ll \tau_{ph\,ph-t}$$
$$\sim \tau_{ph\,ph-T_t} < \tau_{ph-T_t-T_t} \ll \tau_{lU}. \tag{4.151}$$

Thus, knowing the relation between the times we may say some words about the way the relaxation proceeds in such a complicated physical system. According to (4.151) first of all equilibrium in both phonon subsystems is established, and then thermalization of photons occurs due to the connection with transverse phonons in already equilibrium (two penultimate terms of the inequality (4.151)). Since the first (in hierarchic terms) interaction has the form $b^+_{ph\alpha k} b_{ph\alpha' k'} b^+_{tk''}$, where $b^+_{ph\alpha k} (b_{ph\alpha k})$ are operators of creation (annihilation) of a photon with wave vector $\boldsymbol{k}$ and polarization $\boldsymbol{e}_\alpha$, we make the conclusion that in process of connection with transverse phonons the photons acquire the equilibrium temperature T_0 and chemical potential μ_{ph} and stream velocity $\boldsymbol{V}$ not equal to zero. The approach of μ_{ph} and $\boldsymbol{V}$ to zero is provided by the interaction $b^+_{ph\alpha k} b_{tk'} b^+_{tk''}$ (this characterizes the penultimate addend in (4.151)).

Therefore, knowing the complete pattern of relaxation in the dielectric with a pore, we may consider and (moreover) describe any non-equilibrium process to which, in particular, we may refer the theory of heat conduction (Chap. 5).

4.8 Recommendations for Increasing the Electric and Thermal Strength of Porous Structures

In order to give any recommendations for improving the electric and thermal properties of porous media we should again turn to the formulae obtained above and analyze them carefully. We shall start from the electric breakdown formulae (4.35), (4.48), (4.64) and (4.67) in succession. From the first formula it directly follows that increase of E^* is possible at:

(a) decrease of sites of dielectric surface defects (improvement of rolling surface);
(b) decrease of the distance between dipole moments $\boldsymbol{d}_i$ on which independent elastic scattering of electrons takes place;
(c) increase of cross-section of electron scattering $\sigma(T)$ by gaseous phase (say, by the structure impregnation with viscous liquid).

From the dependence (4.48) we conclude:

(d) decrease of a dielectric thickness contributes to increase E^*;

From (4.64) and (4.67) it follows that the increase of E^* is caused by:

(e) increase of fibril dimensions;
(f) increase of thermal diffusion of a fibril structure which is achieved by, for example, adding of alien atoms of, say, zinc or iron into the fibril chemical composition;
(g) simultaneous increase of the fibril diameter $2r_{fl}$ and decrease of their length l_{fl}.

As to the thermal breakdown, from (4.101) and (4.112) it follows that in order to decrease its probability, those substances should be used practically which have

(a) the highest sound velocity c_s;
(b) the greatest structure density ρ;
(c) the smallest thickness δ;
(d) minimum size of in homogeneity region r_{br}.

The latter two points are included in the recommendations for improving the elastic strength. It means that such structures with such characteristics must "work" perfectly under conditions of both thermal and electric breakdown.

5 On Specific Features of Thermal Conduction and Diffusion in Porous Dielectrics

The investigation of the process of thermal conduction has always been very important in practice. It is evident. The application of various substances in thermal-physical devices requires preliminary labor-consuming experimental measurement of their thermal conduction time. If, for example, a thermal insulator is concerned, such substances should be chosen whose thermal diffusion coefficient χ is the lowest, which naturally leads to the largest heating time and consequently to a long period of heat transfer.

Although the measurement of the actual time of thermal conduction has a well adjusted and reliable algorithm, in a number of cases however, especially in complex structures of the cellulose or paper type, besides experimentally found regularities of behavior of χ caused by a number of sample parameters (e.g. its density, environment temperature and others) it is necessary to understand what physical causes underlie this dependence. This question may be answered by theoretical analysis. Anticipating the additional question from a practical point of view (why do we need a theory if such a dependence of χ on the parameters was obtained experimentally), we shall answer it. The matter is that in an overwhelming number of cases the theoretical interpretation of the experiments allows us to make a number of far-reaching *complementary* conclusions (predictions) and forecasts concerning the question how to decrease (or increase) the thermal conductivity coefficient. Indeed, when some parameter of a system is described theoretically, in the formal description of this value there is always a whole number of additionally introduced characteristics (besides the main ones), but they are indispensable.

It is these new characteristics contained in the final formula that allow us to find out and *predict* complementary regularities based on which we can forecast the improvement of the physical properties of a substance and which were not found experimentally. After theoretical prediction the experiment may be repeated in order to find out these new regularities.

To the given type of structure we shall refer all kinds of wood, paper dielectrics, ultra-thin glass fiber, wool felt and others [67]. It may be noted that these dielectrics have strongly abnormal physical properties. For example, the heat capacity behaves in the temperature range 300–450 °K linearly with respect to T (for more details see Sect. 3.2) though it is known (see [11]) that at high temperatures exceeding the Debye temperature θ_D, C_P must not depend on temperature.

The difficulty of solution of our task – analytical calculation of the thermal conductivity coefficient x of porous structures – is, in particular, that in such substances there are at least two heat streams – the phonon stream propagating in the fiber structure and the photon stream caused by radiant heat exchange through the region of free volumes (without taking account of interference between them!).

The problem of the theoretical description of the thermal factors of such media is also real because it is not quite clear what is the real dissipation mechanism of photon and phonon energies in a fiber, or, putting it otherwise, what is the most effective mechanism of interaction in terms of probability. Besides, it is not quite clear what are sound vibrations (phonons) as such in the fiber structure. That is why one of the assumptions which we are making now and which we shall take as basic is the phenomenological introduction of a phonon gas with linear over the wave vector k spectrum. Let us assume, therefore, that the frequency of sound vibrations is $\omega_k = c_{s\nu}k$, where $c_{s\nu}$ is the local sound velocity in the νth fiber. Let the total number of fibers be M.

As to the dissipation mechanism of phonons, which is dominant in the description of heat conduction, traditionally it should be the inelastic mechanism of relaxation. It may be, for example, phonon scattering by structure inhomogeneities, fiber matrix defects, and alien atoms, but in this case an elementary act of interaction must be followed by an energy transfer, though very little.

Before we directly pass over to the solution of our problem we would like to touch, though briefly, upon one more substance very interesting from the investigation point of view. We mean a *magnetic porous dielectric*. We cannot help mentioning these structures, as we are sure of their highly abnormal properties. Indeed, magnetic dielectric media (and porous ones still more!) have always been and still remain very interesting objects of study.

The presence of macroheterogeneities having the shape of small spherical voids (to be concrete we shall discuss spheres) greatly deforms the spectrum of magnetic collective oscillations, also called magnons, and leads to the dependence of the exchange interaction in Heisenberg's model on the porosity of the structure m. If the pore concentration is small, the magnon spectrum displays an ordinary law of spin-wave dispersion, the law being a little bit "spoiled" by macroheterogeneities. If the concentration is high, then, firstly, in such a two-phase system the sound velocity changes (it becomes greatly dependent of the concentration of both phases; see Sect. 3.2), and secondly, the exchange interaction of the spins will represent some composite function of m. When the magnon wavelength λ_m considerably exceeds the pore dimension R_p we may say that the structure is practically either a pure ferro- or antiferromagnet from the point of view of the magnons. Here the spectrum of the quasi-particles is a very good continuous function of the wave vector. If the pores are large in dimension, we cannot say that magnetic oscillations are merely spin waves, for the translation invariance of the lattice is violated and in order to describe the spectrum in this case it is necessary to consider some other model. The simplest example is when all the pores are ordered and are situated at the specified distance d from each other. But even this simple model assumption leads to difficulties of introduction of the magnon spectrum. Here it would be certainly be useful to clear up the dependence of the sound velocity on the concentration of the pores m similarly to what was done, for example, in the second section. Though it is purely philosophical reasoning, for rigorous validation before we introduce the function $c_s = c_s(m)$, it is necessary to calculate it strictly mathematically.

Let us return to our problem. In calculating the heat losses in a fiber we shall assume that the main mechanism of photon scattering is defined by the wavelength from the range $l_{fl} \ll \delta R$, where $\delta R = R_2 - R_1$ is the wall thickness of a fiber and l_{fl} is the extent of the fibrils. Let us choose a working hypothesis. According to Sect. 2.2 we shall suppose that we have at our disposal an ordered fibril structure whose scheme is illustrated in Fig. 5.1. At the same time, as we assumed before, the main mechanism of

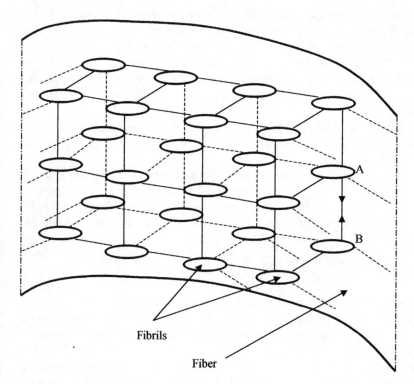

Fibrils

Fiber

Fig. 5.1. Symbolic section of a fibre with schematic pattern of ideal packing of fibrils

photon scattering is determined by the wavelength $\lambda \gg l_F$, and therefore we should agree that the type of packing of fibril structure for photons is *not significant*. This, in turn, means that a photon must be scattered only by fiber density fluctuations. It would be also desirable to understand how fibers behave when heated. Suppose that a fibril A (see Fig. 5.1) is heated and kept at constant temperature T_0. Then due to the contact of fibril A with fibril B the latter will be heated through a "microscopic" region of gaseous interlayer. If the characteristic spatial interval δx in which the temperature $T(x,t)$ significantly changes is much larger than the fibril diameter d_{fl}, then it makes sense to speak only about pure heat conduction, and the problem of calculation of the fiber temperature equalization time is confined to a simple solution of the ordinary Fourier equation of heat conduction.

Since we shall consider photons with the wavelength $\lambda \ll \delta R$, two situations are possible:

$$d_{fl}, l_{fl} \ll \lambda \ll \delta R \tag{5.1}$$

and

$$\lambda \ll d_{fl}, l_{fl} \ll \delta R. \tag{5.2}$$

In the first case radiant heat exchange between contacting parts of fiber surfaces occurs completely due to the heat conduction, but the second case is much more complicated and it should be considered that all the fibers are fixed (from the point of view of photons) at a finite distance from each other, so that heat transfer between them takes place only due to photon exchange. The latter statement shows that the calculations should be done microscopically.

Thus, we shall consider only those photons whose wavelength satisfies the inequality (5.2), and consequently only they are responsible for the radiant heat exchange.

In calculating the thermal conductivity coefficient of the whole dielectric the ensemble of fibrils and their dimension spread should be taken into account. The thermal conductivity of one fiber, which we shall conventionally call local, will be denoted by $\kappa_\nu(m, T)$. Then the average thermal conductivity over the substance volume is

$$\langle \kappa(m, T) \rangle = \frac{\int \Phi(v) \kappa_v(m, T) dv}{\int \Phi(v) dv}, \tag{5.3}$$

where $\Phi(v)$ is the fiber distribution function over volume v.

The fact that we have at once introduced into consideration the temperature T in the function $\kappa_v(m, T)$ means that the time of "internal" relaxation in photon and phonon subsystems at relatively high temperatures $T < \langle \phi_D \rangle$ is small compared to the time of observation; here $\langle \phi_D \rangle$ is a prototype of the Debye temperature in fibrils. In reality it is not only an assumption and the given condition does hold, which may be verified directly.

The formula which we shall need in calculating the "local" coefficient of thermal conductivity is called Kubo's formula [76] and in order to apply it to our problem we should dwell upon its derivation in more detail.

5.1 Kubo's Formula

Because the present monograph is intended for readers not very well acquainted with the methods of theoretical physics, during the first familiarization with the given section it may be omitted and the final formula and dependence may be taken from Figs. 5.6–5.22.

The mathematical apparatus and methods for the calculation of the kinetic parameters of *weakly non-equilibrium* systems were first proposed in 1958 by the Japanese theoretical physicist Kubo. This approach allows us to calculate any known macroscopic coefficients of transfer (of thermal conductivity, viscosity, diffusion, susceptibility, and conduction) in linear approximation for a weak external disturbing factor. It may be, for example, temperature gradient, electric field, external magnetic field, gradient of particle concentration, derivatives from velocity vector components, etc.

The essence of the method is the following. At our disposal we have a quasi-equilibrium density matrix of a large macroscopic system

$$\rho = X^{-1}e^{-H/T-\hat{A}} , \qquad (5.4)$$

where the normalization factor $X = Sp\,e^{-\hat{H}/T}$, Sp means taking a "trace", i.e. the sum over all diagonal states of the system, $\hat{H}$ is the system hamiltonian, and the operator $\hat{A}$ describes the interaction with a *macroscopic* small parameter. We shall adhere to the commonly used notation for operators, and their difference from the conventional c-number we shall denote by the "cap" over the corresponding magnitude.

In the framework of our problem the operator A means the operator of the thermal stream of the form:

$$\hat{A} = \int_V \int_0^{i\infty} \left(\frac{\nabla q}{T}\right) d^3x d\tau = -\int_V \int_0^{i\infty} \left(\frac{\nabla q}{T}\right) d^2x d\tau , \qquad (5.5)$$

where τ is imaginary time ($\tau = it$) and q is the operator of the microscopic thermal stream.

Any average observed value within the density matrix method is determined by the relation (e.g. see [94]):

$$\langle B \rangle = Sp\,(\rho B) , \qquad (5.6)$$

where the angle brackets mean averaging over the equilibrium density matrix, which is identical to taking a "trace" by the corresponding quantum numbers.

In order to use the formula (5.6) it is necessary to calculate preliminarily the density matrix in a linear manner over the operator $\hat{A}$ approximation.

Note that below the "cap" at the Hamiltonian we shall cut off. From the equation of "motion" for ρ, according to which

$$\hbar \frac{\partial \rho}{\partial \tau} = [\rho, H + V],$$

where the operator $V(\tau) = \partial \hat{A}(\tau)/\partial \tau$, and the square brackets mean commutation, in the V-linear approximation we find that

$$\rho \cong \rho_0 - \int\limits_{-\infty}^{\tau} [\rho, V] \frac{\mathrm{d}\tau}{\hbar}, \tag{5.7}$$

where ρ is the density matrix (5.4) at $\hat{A} = 0$. Substituting now (5.7) into (5.6) we get

$$\langle B \rangle = Sp(\rho_0 B) - Sp \int\limits_{-\infty}^{\tau} ([\rho_0, V(\tau)]B)\mathrm{d}\tau = B_0 - Sp \int\limits_{-\infty}^{\tau} (\rho_0[B, V(\tau)])\mathrm{d}\tau$$

$$= B_0 - Sp\rho_0 \int\limits_{-\infty}^{\tau} \left[B, \frac{\partial \hat{A}(\tau)}{\partial \tau} \right] \mathrm{d}\tau = B_0 + Sp\rho_0[B, \hat{A}(\tau)], \tag{5.8}$$

where $B_0 = \langle B \rangle_0 = Sp(\rho_0 B)$ and we take $\hat{A}(-\infty) = 0$.

The general formula (5.8) allows us to calculate any kinetic coefficient of interest. For our problem, at $B = q$ we have taking account of (5.5) that

$$\langle q \rangle = - \int\limits_{V} \int\limits_{0}^{i\infty} \langle [q(0,0), q(x,\tau)] \rangle_0 \frac{\mathrm{d}^3 x \mathrm{d}\tau \nabla T}{T^2} \tag{5.9}$$

or as the average thermal stream according to Fourier's law is $\langle q \rangle = \kappa \nabla T$, for the thermal conductivity tensor from (5.9) we find

$$\kappa_{\alpha\beta} = \int\limits_{V} \int\limits_{0}^{i\infty} \langle [q(0,0), q(x,\tau)] \rangle_0 \frac{\mathrm{d}^3 x \mathrm{d}\tau}{T^2}. \tag{5.10}$$

For our problem it will be more convenient to use the thermal stream related to a unit volume . Taking therefore $q = q/V$ and introducing the factor $e^{i\varepsilon\tau}$, which provides convergence of the integral in "time", we find

$$\kappa_{\alpha\beta} = \lim_{\varepsilon \to 0} \int\limits_{V} \int\limits_{0}^{i\infty} e^{i\varepsilon\tau} \langle [q_\alpha(0,0), q_\beta(x,\tau)] \rangle_0 \frac{\mathrm{d}^3 x \mathrm{d}\tau}{T^2}. \tag{5.11}$$

The latter formula will be used to calculate the functional dependence of the local thermal conductivity $\kappa_v(m, T)$.

5.2 Green's Functions and Correlators

Despite the fact that the derivation of Kubo's formula occupies only two lines, its application in certain particular cases including the calculation of x faces rather serious difficulties of a purely mathematical character (see [95–101]). Nevertheless, it does not cause pessimism but requires more careful and accurate calculating.

As was mentioned above, taking account of the radiant transfer of heat in pores there are at least two streams: one is connected with photons and the other with phonons. Between them there is an interaction. Kubo's formula facilitates the account of these two streams. Indeed, let us very slightly change the formula (5.11) and write it in the following form:

$$\kappa_{\alpha\beta} = (VT^2)^{-1} \lim_{\varepsilon \Rightarrow 0} \int_V \int_0^{i\infty} e^{i\varepsilon(\tau-\tau')} \langle [\delta q_\alpha(0), \delta q_\beta(\tau - \tau')] \rangle_0 \mathrm{d}(\tau - \tau'), \quad (5.12)$$

where the Greek indices denote x, y, z, V is the volume of the whole body and the operator of the thermal stream deviation from the equilibrium stream is (at $\tau = 0$)

$$\delta q_\alpha(\tau) = \sum_{j=1}^{2} [q_{j\alpha}(\tau) - q_{j\alpha}(0)]$$

$$= N^{-1} \sum_j \sum_k \hbar\omega_{jk} v_{\alpha jk} [n_{jk}(\tau) - \langle n_{jk} \rangle], \quad (5.13)$$

where the operator of the number of particles is written down in Heisenberg's representation [86]

$$n_{jk}(\tau) = \exp\{-\int H_{j\mathrm{int}}(\tau)\mathrm{d}\tau\} n_{jk} \exp\{\int H_{j\mathrm{int}}(\tau)\mathrm{d}\tau\}. \quad (5.14)$$

N is the number of lattice sites in the solid matrix, $H_{j\mathrm{int}}$ is the hamiltonian of the interaction of the jth subsystem (we mean either photons or phonons).

Their equilibrium distributions are described by common Bose functions:

$$\langle n_{jk} \rangle = \left(\exp\left\{ \frac{\hbar\omega_{jk}}{T} \right\} - 1 \right)^{-1}, \quad (5.15)$$

$$\omega_{jk} = \begin{cases} \omega_{1k\nu} & \text{for phonons in the } \nu\text{th fiber} \\ \omega_{2k\nu} & \text{for photons}. \end{cases} \quad (5.16)$$

Taking account of these formulae we shall rewrite the relation (4.12) in another way. Introducing the porosity $m = V_p/V$ and assuming that the operator of the thermal stream q is the additive magnitude composed of two components, the photon stream and the phonon stream, in the isotropic case we have

$$\kappa(m, T) = (VT^2)^{-1}\hbar^2 \lim \left[\int_0^{i\infty} d\tau \{ (1-m)^2 \sum\sum c_s^4 k_1 k_2 B_1(\boldsymbol{k}_1, \boldsymbol{k}_2, \tau) \right.$$

$$+ m(1-m) \sum\sum c^2 c_s^2 k_1 k_2 B_2(\boldsymbol{k}_1, \boldsymbol{k}_2, \tau)$$

$$+ m(1-m) \sum\sum c^2 c_s^2 k_1 k_2 B_3(\boldsymbol{k}_1, \boldsymbol{k}_2, \tau)$$

$$\left. + m^2 \sum\sum c^4 k_1 k_2 B_4(\boldsymbol{k}_1, \boldsymbol{k}_2, \tau) \} e^{-i\varepsilon\tau} \right], \tag{5.17}$$

where

$$B_1(\boldsymbol{k}, \boldsymbol{k}', \tau) = \langle (b_k^+ b_k - \langle n_{1k}\rangle)[b_{k'}^+ b_{k'}](\tau)\rangle, \tag{5.18}$$

$$B_2(\boldsymbol{k}, \boldsymbol{k}', \tau) = \langle (c_k^+ c_k - \langle n_{2k}\rangle)[b_{k'}^+ b_{k'}](\tau)\rangle, \tag{5.19}$$

$$B_3(\boldsymbol{k}, \boldsymbol{k}', \tau) = \langle (b_k^+ b_k - \langle n_{1k}\rangle)[c_{k'}^+ c_{k'}](\tau)\rangle, \tag{5.20}$$

$$B_4(\boldsymbol{k}, \boldsymbol{k}', \tau) = \langle (c_k^+ c_k - \langle n_{2k}\rangle)[c_{k'}^+ c_{k'}](\tau)\rangle. \tag{5.21}$$

According to the definition (5.16)

$$\langle n_{k\nu}\rangle = \langle n_{1k\nu}\rangle \quad \text{and} \quad \langle N_{k\mu}\rangle = \langle n_{2k\mu}\rangle,$$

where ν and μ correspond to the polarization of phonons and photons respectively.

As to the "time" dependence of operators of the number of particles n, they are determined by the relations

$$[b_k^+ b_k](\tau) = \exp\left\{ -\int H_{1\text{int}}(\tau)d\tau \right\} b_k^+ b_k \exp\left\{ \int H_{1\text{int}}(\tau)d\tau \right\}, \tag{5.22}$$

$$[c_k^+ c_k](\tau) = \exp\left\{ -\int H_{2\text{int}}(\tau)d\tau \right\} c_k^+ c_k \exp\left\{ \int H_{2\text{int}}(\tau)d\tau \right\}, \tag{5.23}$$

where $c_{k\mu}^+ (c_{k\mu})$ are operators of creation (annihilation) of photons and $b_{k\nu}^+ (b_{k\nu})$ are operators of creation (annihilation) of photons in the λth fiber.

In order to calculate the correlators (5.18)–(5.21) it is necessary to introduce additionally the following Green's functions:

$$D_{1kk'}(\tau - \tau') = \langle T_\tau | b_k^+(\tau)b_k(\tau')(c_{k'}^+ c_{k'} - \langle n_{2k'}\rangle)|\rangle, \tag{5.24}$$

$$D_{1kk'k''}*(\tau - \tau') = \langle T_\tau | b_k^+(\tau)b_{k''}(\tau')(c_{k'}^+ c_{k'} - \langle n_{2k'}\rangle)|\rangle, \tag{5.25}$$

$$D_{2kk'}(\tau - \tau') = \langle T_\tau | c_k^+(\tau)c_k(\tau')(b_{k'}^+ b_{k'} - \langle n_{1k'}\rangle)|\rangle, \tag{5.26}$$

$$D_{2kk'k''}*(\tau - \tau') = \langle T_\tau | c_k^+(\tau)c_{k''}(\tau')(b_{k'}^+ b_{k'} - \langle n_{1k'}\rangle)|\rangle, \tag{5.27}$$

$$D_{1k}^{**}(\tau - \tau') = \langle T_\tau | b_k^+(\tau)b_k(\tau')b_k^+ b_k|\rangle, \tag{5.28}$$

$$D_{1kk'}^{**}(\tau - \tau') = \langle T_\tau | b_k^+(\tau)b_{k'}(\tau')b_{k'}^+ b_{k'}|\rangle, \tag{5.29}$$

$$D_{2k}^{**}(\tau - \tau') = \langle T_\tau | c_k^+(\tau) c_k(\tau') c_k^+ c_k | \rangle \,, \tag{5.30}$$

$$D_{2kk'}^{**}(\tau - \tau') = \langle T_\tau | c_k^+(\tau) c_{k'}(\tau') c_{k'}^+ c_{k'} | \rangle \,. \tag{5.31}$$

T_τ is the operator of chronological ordering which arranges standing behind it operators of creation (annihilation) in order of decreasing "time".

Note, that all of them are written done in Heisenberg's representation and their connection with the interaction representation is as usual [86]

$$b_k(\tau) = \exp\left\{-\int H_{1\mathrm{int}}(\tau)\mathrm{d}\tau\right\} b_{0jk}(\tau) \exp\left\{\int H_{1\mathrm{int}}(\tau)\mathrm{d}\tau\right\} \,,$$

$$c_k(\tau) = \exp\left\{-\int H_{2\mathrm{int}}(\tau)\mathrm{d}\tau\right\} c_{0jk}(\tau) \exp\left\{\int H_{2\mathrm{int}}(\tau)\mathrm{d}\tau\right\} \,, \tag{5.32}$$

and

$$b_{0k}(\tau) = \exp\left\{-H_0(\tau)\right\} b_k \exp\left\{H_0(\tau)\right\} = \exp\{-\omega_{1k\nu}\tau\}b_k \,,$$

$$c_{0k}(\tau) = \exp\left\{-H_0(\tau)\right\} c_k \exp\left\{H_0(\tau)\right\} = \exp\{-\omega_{2k\mu}\tau\}c_k \,. \tag{5.33}$$

By means of Eqs. (5.32) and (5.33) it is easy to find the correspondence between the correlators (5.18)–(5.21) and the Green's functions (5.24)–(5.31). For example, we have that

$$B_{1k}(\tau - \tau') = D_{1k}^{**}(\tau - \tau') \left| \begin{array}{l} \exp\{\omega_{lk}(\tau - \tau')\} \,, \\ \tau > \tau' \end{array} \right. \tag{5.34}$$

$$B_{2kk'}(\tau - \tau') = D_{1kk'}^{*}(\tau - \tau') \left| \begin{array}{l} \exp\{\omega_{lk}(\tau - \tau')\} \,, \\ \tau > \tau' \end{array} \right. \tag{5.35}$$

5.3 Calculation of the Thermal Conductivity Coefficient

After the introduction of the correlators and the Green's functions necessary for further calculations we may proceed directly to our final purpose, namely, to the calculation of the thermal conductivity coefficient of greatly disordered pores structures. Let us begin our calculation by determining the dependence of the function $B_{2\nu}(k, k', \tau)$. In order to calculate it we should first find the equations of "motion" for the Green's functions (5.24) and (5.26). For it we shall need the explicit equations of the interaction hamiltonians $H_{1,2\mathrm{int}}$. For the hamiltonian of the interaction of phonons with fiber matrix defects we have

$$H_{1\mathrm{int}} = \sum\sum\sum A(k, k') b_k^+ b_{k'} e^{i(k-k')x} \,, \tag{5.36}$$

where the summation is performed over all the sites of local arrangement of defects i and also over all wave vectors. The process amplitude is

$$A(k, k') = \gamma_1 \langle \theta_D \rangle \varepsilon \hbar \frac{(e_{1j}k)(e_{2j}k')}{(2\rho v(\omega_k \omega_{k'})^{1/2})} \,, \tag{5.37}$$

where γ_1 is the constant of phonon relation with the site defects, v is the volume of one fiber, e_{1j} is the polarization vector, $\varepsilon = \delta r / r_0 = (r_0 - \langle r \rangle)/r_0$ is the deformation parameter characterizing the extent of changing of the main distances by alien atoms, $\delta r_0 = (r_0 - \langle r \rangle)$, and r_0 is the initial distance in the non-deformed structure and $\langle r \rangle$ in the deformed one.

Now about the energy operator of photon interactions with fluctuations of the fiber structure density. For its calculation it is necessary to write down the following invariant Hamiltonian:

$$H_{2\mathrm{int}} = - \left(\frac{\gamma_2}{2\rho} \right) \int (\boldsymbol{E} \mathrm{grad}\rho) \, \mathrm{div}\, \boldsymbol{E} \mathrm{d}^3\boldsymbol{x} \,, \tag{5.38}$$

where γ_2 is the coupling constant having dimensions $[\mathrm{cm}^2]$. $\boldsymbol{E}$ is the electric field operator. Holding that the photon wavelength satisfies the equality

$$R_0 \,, \quad \langle r \rangle \ll \lambda \ll b \,, \tag{5.39}$$

we shall write the operator $\boldsymbol{E}$ as a representation of secondary quantization over operators. As $\boldsymbol{E} = -c^{-1}\partial A/\partial t$, and the operator wave function of a free photon is

$$\boldsymbol{A} = -c \sum_k \sum_j \left(\frac{4\pi\hbar}{\omega_{2k}V} \right)^{1/2} e_{2j}[c_{kj}(t) + c^{+}_{-kj}(t)] \exp(\mathrm{i}\boldsymbol{kx}) \,,$$

we have

$$\boldsymbol{E} = \mathrm{i} \sum_k \sum_j \left(\left(\frac{4\pi\hbar}{\omega_{2k}V} \right)^{\frac{1}{2}} e_{2j}[c_{kj}(t) - c^{+}_{-kj}(t)] \exp(\mathrm{i}\boldsymbol{kx}) \right) \,, \tag{5.40}$$

where $e_{2\mathrm{i}}$ is the vector of photon polarization;

$$\begin{cases} c_{kj}(t) = c_{kj} \exp(-\mathrm{i}\omega_{kj}t) \quad \text{and} \\ c^{+}_{kj}(t) = c^{+}_{kj} \exp(\mathrm{i}\omega_{kj}t) \,. \end{cases}$$

Substituting this expression into the Hamiltonian (5.38) and taking into account the expansion of the density operator into a Fourier series

$$\delta\rho = \sum \delta\rho_k e^{-\mathrm{i}kx}$$

we get the sought for interaction of photons with density fluctuations as

$$H_{2\mathrm{int}} = \sum\sum \psi(\boldsymbol{q},\boldsymbol{k},\boldsymbol{k}')\delta\rho_q c^{+}_k c_{k'} \Delta(\boldsymbol{q}+\boldsymbol{k}-\boldsymbol{k}') \,, \tag{5.41}$$

where the amplitude of the process is

$$\psi(\boldsymbol{q},\boldsymbol{k},\boldsymbol{k}') = \left(\frac{2\pi\gamma_2\hbar}{\rho_0} \right) (\boldsymbol{k}e_{2j})(\boldsymbol{k}'e_{2j})(\omega_{2k}\omega_{2k'})^{1/2} \,. \tag{5.42}$$

Knowing, thus, the Hamiltonians (5.36) and (5.41) we can begin deriving equations of "motion" for the Green's function. Let us choose for this purpose, for example, the function (5.28) and represent it by means of Heaviside's functions in the form:

$$D^{**}_{1kk}(\tau - \tau') = \theta(\tau - \tau')\left\langle |T_\tau| b^+_k(\tau) b_k(\tau') b^+_k b_k| \right\rangle$$
$$+\theta(\tau' - \tau)\left\langle |T_\tau| b^+_k(\tau') b_k(\tau) b^+_k b_k| \right\rangle . \tag{5.43}$$

Differentiating this equation with respect to τ_1 and using the equation of "motion" for Bose operators

$$\frac{\partial b_{k\nu}}{\partial \tau} = [H_{\text{int}}, b_{k\nu}] ,$$

where $H_\nu = H_{0\nu} + H_{1\text{int}}$ and $H_{0\nu} = \sum_k \omega_{1k\nu} b^+_{k\nu} b_{k\nu}$, we get

$$\frac{\partial D^{**}_{1kk}}{\partial \tau} = (1 + \langle n_{1k}\rangle)\delta(\tau - \tau') - \omega_{1k} D^{**}_{1kk} - \sum_{k'} A_{kk'} D_{1k'k} , \tag{5.44}$$

where the function D_{1k_1k} is determined by (5.25). In an analogous way we may obtain the equation

$$\frac{\partial D^{**}_{1kk'}}{\partial \tau} = -\omega_{1k} D^{**}_{1kk'} - \sum_{k''} A_{kk''} D_{1k''k'} . \tag{5.45}$$

Making in (5.44) and (5.45) a Fourier transformation over time we find the following system:

$$\begin{cases} D^{**}_{1kk}(i\omega_s - \omega_{1k} = (1 + \langle n_{1k}\rangle) + \sum_{k'} A_{kk'} D_{1k'k} , \\ D^{**}_{1kk'}(i\omega_s - \omega_{1k'}) = \sum_{k''} A_{kk''} D_{1k''k} , \end{cases} \tag{5.46}$$

where $D^{**}_{1kk}(\omega_s) = 0.5 \int D^{**}_{1kk}(\tau) \exp\{i\omega_s\tau\}d\tau$, $\omega_s = 2\pi Ts$, $s = 1, 2, 3, \ldots$

In order to uncouple the obtained equations we get

$$D^{**}_{1kk'}(\omega_s) = \Delta_{kk'} D_{1kk}(\omega_s) , \tag{5.47}$$

where

$$\Delta_{kk'} = \begin{cases} 1 \text{ at } k = k' \\ 0 \text{ at } k \neq k' . \end{cases}$$

Taking this into account, the solutions of the equations are

$$D^{**}_{1kk}(\omega_s) = \frac{(1 + \langle n_{1k}\rangle)}{[i\omega_s - \omega_{1k} - \sum_{k'} |A_{kk'}|^2/(i\omega_s - \omega_{1k'})} . \tag{5.48}$$

And quite analogously we get

$$D^{**}_{1kk'}(\omega_s) = \frac{(1 + \langle n_{1k'}\rangle)}{[i\omega_s - \omega_{1k} - \sum_{k'} |A_{kk'}|^2/(i\omega_s - \omega_{1k'})} . \tag{5.49}$$

Such calculations allow us to calculate the photon functions of Green for the interaction of photons with density fluctuations of a fiber described by dependence (5.41). For example,

$$D_{2kk'}^{**}(\omega_s) = \frac{(1 + \langle n_{2k'} \rangle)}{[i\omega_s - \omega_{2k} - \sum_{k,q} |\psi(k', k, q)|^2 \langle \delta\rho_q \delta\rho_{-q} \rangle / (i\omega_s - \omega_{2k})}. \quad (5.50)$$

Making in relations (5.48)–(5.50) an inverse Fourier transformation according to the formula

$$D_k(x) = 2\pi T \sum_s D_k(\omega_s) \exp\{-i\omega_s x\}, \quad (5.51)$$

we get at $\tau_1 > \tau_2 (\tau > 0)$

$$D_{1kk'}^{**}|_{\tau>0, k=k'} = (1 + \langle n_{1k'} \rangle) \langle n_{1k'} \rangle \exp\{-(\omega_{1k} - i\gamma_{1k})\tau\}, \quad (5.52)$$

where the attenuation

$$\gamma_{1k} = 2\pi\hbar^{-2} \sum_{k'} |A(k, k')|^2 \delta(\omega_{1k} - \omega_{1k'}), \quad (5.53)$$

$$D_{2kk'}^{**}(\tau)|_{\tau>0, k=k'} = \langle n_{2k} \rangle \exp\{-(\omega_{2k} - i\gamma_{2k})\tau\}, \quad (5.54)$$

where the attenuation of the photons is

$$\gamma_{2k} = 2\pi\hbar^{-2} \sum_{q,k'} |\psi(k', k, q)|^2 \langle \delta\rho_q \delta\rho_{-q} \rangle \delta(\omega_{2k} - \omega_{2k}') \Delta(q + k - k'). \,(5.55)$$

Let us calculate both relations (5.53) and (5.55). We shall start with (5.53).

According to (5.37), after passing in (5.53) from summing to integration with respect to k we find

$$\gamma_{1k} = \gamma_{10} + \gamma_1^2 c_d \langle \theta_D \rangle^2 \frac{\varepsilon^2 k^4}{60\pi c_s^3 \rho_f^2 v}, \quad (5.56)$$

where the first addend in the obtained relation describes the "saturation" of phonon attenuation occurring in the region of low temperatures (the so-called Knudsen limit). It means that $\gamma_{10} = c_{sv}/b$. Note that b is the longitudinal dimension of a fiber, $c_d = N_d/N$ is the concentration of defects in a fiber, and N_d is the number of defects.

To calculate the dependence γ_{2k} we need the density correlator $\sum q^2 \langle \delta\rho_q \delta\rho_{-q} \rangle$. We shall show how to find it. Let us express the energy density in a fiber as

$$\varepsilon_v(\rho) = \varepsilon_v(\rho_0 + d^* \nabla\rho) = \varepsilon_v(\rho_0) + \frac{P_f M \, d^* \nabla\rho}{\rho_0^2}$$

$$+ 0.5 \left(\frac{\partial}{\partial\rho_0} \right) \left(\frac{P_f M}{\rho_0^2} \right) (d^* \nabla\rho)^2 + \dots,$$

where P_f is the "pressure" in a unit of fiber volume, d^* is the some size, and M is the mass of the fiber. It means that the entire energy (Hamiltonian) is

$$H(\rho) = \int_{v_f} \varepsilon_v dv = \left(\frac{P_f M d^{*2}}{3\rho_0^3}\right) \sum_q q^2 \delta\rho_q \delta\rho_{-q},\qquad(5.57)$$

and therefore, according to the general theory of fluctuation phenomena [11], the sought for correlator is

$$\sum_q q^2 \delta\rho_q \delta\rho_{-q} = \frac{3\rho_0^3 T}{2PMd^{*2}},\qquad(5.58)$$

where P is the "pressure" in the in a fiber.

Taking account of the relation (5.58) and the scattering amplitude (5.42) from (5.55) we find the attenuation coefficient of interest to us as

$$\gamma_{2k} = \frac{\gamma_2^2 \rho_0 c \langle a\rangle^5 k^4 T}{10 P M d^{*2}}.\qquad(5.59)$$

Besides photon scattering in the fibers it also occurs in the free volume $v_p (v_p > b^3)$. This mechanism may be accounted for if we use the formula for the extinction coefficient h, which is valid and applicable to our case if the following conditions hold:

$$l_M \ll \lambda \ll b,\qquad(5.60)$$

where l_M is the free path of molecules in the gaseous phase. According, for example, to [37] we have

$$h = \left(\frac{v_p}{6\pi}\right)\left(\frac{\omega}{c}\right)^4 \langle \delta\varepsilon^2\rangle_v$$

or because $\gamma = hc$, then

$$\gamma_{2k}^l = \left(\frac{2}{3\pi N_0}\right) ck^4 \left(\varepsilon_0^{1/2} - 1\right)^2 v_p,\qquad(5.61)$$

where ε_0 is the dielectric permeability of the gas, N_0 is the total number of molecules in the specified free volume. Consequently, the attenuation coefficient sought for is

$$\begin{aligned}\gamma_{2k}^{(ef)} &= \gamma_{2k} + \gamma_{2k}^l + \gamma_{20}\\ &= \frac{\gamma_2^2 \rho_0 c \langle a\rangle^5 k^4 T}{10 P M d^{*2}} + \left(\frac{2}{3\pi N_0}\right) ck^4 \left(\varepsilon_0^{1/2} - 1\right)^2 v_p + \gamma_{20},\end{aligned}\qquad(5.62)$$

where $\gamma_{20} = c/L_p$, L_p is the linear dimension of the pores.

In order to calculate the correlators (5.19) and (5.20) we also should know the interaction between the photon and phonon subsystems. The corresponding invariant hamiltonian may be presented, for example, as

$$H_{v\text{int}}^{(12)} = \gamma_3 \int E_i E_k \boldsymbol{u}_{ik} d^3 x,\qquad(5.63)$$

where γ_3 is a dimensionless coupling constant and $v_{f\nu}$ is the volume of νth fiber.

Substituting in (5.63) the Eq. (5.40) and the operator

$$u_{ik} = 0.5 \left(\frac{\partial u_i}{\partial x_k} + \frac{\partial u_k}{\partial x_i} \right) = i \sum \left(\frac{\hbar}{2\rho_{f\nu}\omega_{1k}v_{f\nu}} \right)^{1/2} (e_{1j\nu}\boldsymbol{k})(b_{jk}^+ - b_{j-k})e^{i\boldsymbol{kx}}$$

we find

$$H_{\nu\mathrm{int}}^{(12)} = \sum_{qq'k} \sum_j \sum_{\mu\mu'} \psi_{j\mu\mu'}^{(12)}(\boldsymbol{q}, \boldsymbol{q}', \boldsymbol{k})(c_{q\mu}^+ - c_{-q\mu})(c_{q'\mu'}^+ - c_{-q'\mu'})$$
$$\times (b_{kj}^+ - b_{-kj})\Delta(\boldsymbol{q} + \boldsymbol{q} - \boldsymbol{k}),$$

(5.64)

where the scattering amplitude

$$\psi_{j\mu\mu'}^{(12)}(\boldsymbol{q}, \boldsymbol{q}', \boldsymbol{k}) = 4\pi i v_{f\nu}\hbar\gamma_3(e_\mu\boldsymbol{q})(e_{\mu'}\boldsymbol{q}') \left(\frac{\hbar\omega_{2q\mu}\omega_{2q'\mu'}}{2\rho_{f\nu}v_f\omega_{1jk}} \right)^{1/2}.$$

(5.65)

Further, we shall also need equations for the attenuation coefficients of photons by phonons and phonons by photons. The relevant calculations are quite similar to those given, for example, in [39] and as a result of the first Born approximation for the interaction we obtain

$$\gamma_{21k} = \gamma_{\mathrm{kphot-phon}} = 64\pi\hbar\gamma_3^2(1 - m)^2 ck^5 \frac{(1 + 2\langle n_{1k}\rangle)}{27c_s\rho_{f\nu}},$$

(5.66)

$$\gamma_{12k} = \gamma_{\mathrm{kphon-phot}} = 2\pi\gamma_3^2(1 - m) \left(\frac{c}{c_{s\nu}} \right)^3 k^4 T \frac{\exp(-\hbar\omega_{1k}/T)}{9\rho_{f\nu}c_{s\nu}},$$

(5.67)

$$\gamma_{12k} = \gamma_{\mathrm{kphon-phot}} = \frac{4\pi\gamma_3^2(1 - m)(c_{s\nu}/c)^3 k^3 T^2}{\hbar\rho_{f\nu}cc_{s\nu}}.$$

(5.68)

Note, that the formula (5.67) is valid only on condition that

$$\langle \theta_D \rangle < T < \frac{\hbar c}{2l_{f\,\mathrm{max}}},$$

(5.69)

where $l_{f\,\mathrm{max}}$ is the maximum fiber length, and the formula (5.68) is valid at

$$T > \frac{\hbar c}{2l_{f\,\mathrm{max}}}.$$

(5.70)

Now we have at our disposal all the parameters and dependences necessary for further calculations.

From formulae (5.19), (5.20), (5.22), and (5.25) there are the following correspondences:

$$B_2(\boldsymbol{k}, \boldsymbol{k}', \tau) = \lim_{E_1 \Rightarrow 0} D_{1kk'}(\tau) \Big|_{\tau > 0},$$

(5.71)

$$B_3(\boldsymbol{k}, \boldsymbol{k}', \tau) = \lim_{E_2 \Rightarrow 0} D_{2kk'}(\tau)\Big|_{\tau>0} , \tag{5.72}$$

where the formally introduced primer energies $E_{1,2}$ are given by the relations

$$H_0 = E_1 b_k^+ b_k + E_2 c_k^+ c_k . \tag{5.73}$$

Let us calculate, for example, the functions $D_{1kk'\nu}(\tau)$. For it we shall introduce the temperature matrix of density:

$$\sigma(T) = T_\tau \exp\{- \int_0^\beta H_{\text{int}} d\tau\}, \tag{5.74}$$

where $\beta = 1/T$. Then

$$D_{1k,k'}(\tau - \tau') = \frac{\langle T_\tau | b_k^+(\tau) b_{k'}(\tau')(c_k^+ c_k - \langle n_{2k} \rangle)\sigma(T)| \rangle_0}{\langle \sigma(T) \rangle} . \tag{5.75}$$

Using further the interaction hamiltonian, defined by the formula (5.64), we can make an expansion in terms of perturbation theory and obtain the set of plots shown in Fig. 5.2. the "Common" Green's functions in Fig. 5.2 are given by equations

$$\alpha_k(\tau - \tau') = \langle T_\tau | c_k^+(\tau) c_k(\tau')| \rangle , \tag{5.76}$$

$$\beta_k(\tau - \tau') = \langle T_\tau | b_k^+(\tau) b_k(\tau')| \rangle . \tag{5.77}$$

By means of these Green's functions the mixed function $D_{1kk'}$ may be plotted in the form of "step" diagrams, as shown in Fig. 5.3.

Summing the step diagrams we find

$$D_{1qk}(\omega_s) = \frac{2\pi T \sum_{s'} g_{qk}(\omega_{s'})}{(1 - \Delta(\omega_{s'}))} , \tag{5.78}$$

where

$$g_{qk}(\omega_s) = \sum_{q'} |\psi(\boldsymbol{q}, \boldsymbol{q}', \boldsymbol{k})|^2 \alpha_q^2(\omega_{s'}) \beta_k^2(\omega_s) \alpha_{q'}(\omega_{s'} - \omega_s) , \tag{5.79}$$

$$\Delta(\omega_s) = 2\pi T \sum_{s'} \sum_{k'} |\psi(\boldsymbol{q}, \boldsymbol{k}, \boldsymbol{k}')|^2$$
$$\times \beta_k(\omega_{s''} + \omega_{s'} - \omega_s) \alpha_q(\omega_{s'} - \omega_s) \alpha_{k'}(\omega_{s'}) . \tag{5.80}$$

Making a summation over s' (this method is described in [86]) and over $\boldsymbol{k}'$ and $\boldsymbol{q}'$ we find

$$D_{1qk}(\omega_s) = \frac{(1 + \langle n_k \rangle)\langle n_k \rangle \gamma_{qk}^*}{(i\omega_s - E_1 + i\Gamma_{12k})^2} , \tag{5.81}$$

where the magnitude

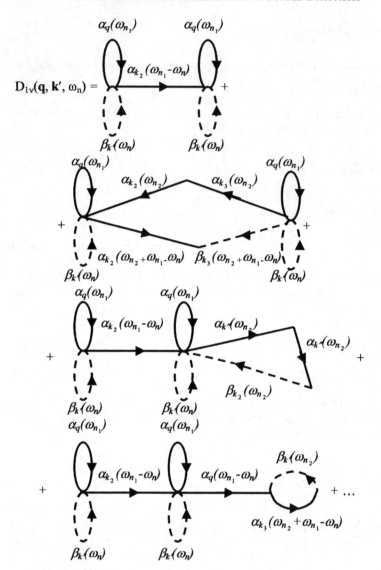

Fig. 5.2. Set of diagrams characterizing mixed thermal streams

$$\gamma_{qk}^* = 2\pi\hbar^{-2} \sum_{q'} |\psi(\boldsymbol{q}, \boldsymbol{q}', \boldsymbol{k})|^2 (\langle n_{2q} \rangle - \langle n_{2q'} \rangle)$$

$$\times \Delta(\boldsymbol{q} - \boldsymbol{q}' - \boldsymbol{k})\delta(\omega_{2q} - \omega_{2q'} - \omega_{1k}) \tag{5.82}$$

and Γ_{12k} is given by formulae (5.67) and (5.68). Passing now into the τ-representation we get

$$D_{1v}(\mathbf{q}, \mathbf{k}', \omega_n) = \qquad + $$

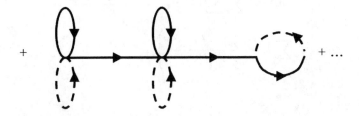

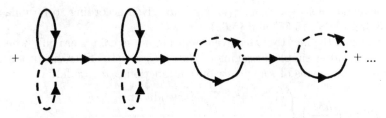

Fig. 5.3. Set of diagrams of "step" type making the main contribution to the direct energy function of the mixed correlate

$$D_{1qk} = -i \lim_{E_1 \Rightarrow 0} (1 + \langle n_{1k} \rangle) \langle n_{1k} \rangle \gamma_{qk}^* \left(\frac{\partial}{\partial \gamma_{12k}} \right) [\exp(-E_1\tau - i\gamma_{12k}\tau)] \quad (5.83)$$

and analogously we find that

$$D_{2qk} = -i \lim_{E_2 \Rightarrow 0} (1 + \langle n_{2k} \rangle) \langle n_{2k} \rangle \gamma_{qk}^{**} \left(\frac{\partial}{\partial \gamma_{21k}} \right) [\exp\{-E_2\tau - i\gamma_{21k}\tau\}], \quad (5.84)$$

where the function γ_{qk}^{**} is given by

$$\gamma_{qk}^{**} = 2\pi\hbar^{-2} \sum_{q'} |\psi(\mathbf{q}, \mathbf{q}', \mathbf{k})|^2 (\langle n_{1q} \rangle - \langle n_{1q'} \rangle)$$

$$\times \Delta(\mathbf{q} - \mathbf{q}' - \mathbf{k}) \delta(\omega_{2q} - \omega_{2q'} - \omega_{1k}). \quad (5.85)$$

In Eqs. (5.83) and (5.84) notations $\gamma_{21k} = \gamma_{\text{kphot-phon}}$ and $\gamma_{12k} = \gamma_{\text{kphon-phot}}$ are introduced for brevity.

Substitution of correlators $B_{1,2,3,4}(\mathbf{k}, \mathbf{k}'\tau)$ taking account of relations (5.34), (5.35) and Eqs. (5.52), (5.54), (5.83) and (5.84) into the formula (5.17) after integration with respect to the "time" interval $\tau_1 - \tau_2$ allows us to get the formula for local thermal conductivity:

$$
\begin{aligned}
\kappa_v(m,T) = (VT^2)^{-1}\hbar^2 \Bigg\{ & (1-m)^2 c_s^4 \sum_k k^2 \frac{(\langle n_{1k}^2 \rangle - \langle n_{1k} \rangle^2)}{\gamma_{1k}} \\
& + m(1-m)c^2 c_s^2 \sum_k \sum_{k'} kk' \langle n_{1k'} \rangle \frac{(1+\langle n_{1k'} \rangle)\gamma_{kk'}^*}{\gamma_{12k'}^{*2}} \\
& + m(1-m)c^2 c_s^2 \sum_k \sum_{kk'} kk' \langle n_{2k'} \rangle \frac{(1+\langle n_{2k'} \rangle)\gamma_{kk'}^{**}}{\gamma_{21k'}^{*2}} \\
& + m^2 c^4 \sum_k k^2 \frac{(\langle n_{2k}^2 \rangle - \langle n_{2k} \rangle^2)}{\gamma_{2k}^{(ef)}} \Bigg\},
\end{aligned}
\tag{5.86}
$$

where

$$
\begin{cases}
\gamma_{12k}^* = \gamma_{12k} + \gamma_{1k}, \\
\gamma_{21k}^* = \gamma_{21k} + \gamma_{2k}^{(ef)}
\end{cases}
\tag{5.87}
$$

and the attenuations γ_{21k}, γ_{12k}, γ_{1k}, $\gamma_{2k}^{(ef)}$ are given according to formulae (5.66), (5.67), (5.68), (5.56) and (5.62).

At last, using the fact that the given sums are not equal to zero only when $k = k_1$, $\nu = \nu_1$, after simple averaging of $H_v(m,T)$ over the fiber dimensions, which is done by means of some distribution function $\varphi(\nu)$, we find

$$
\begin{aligned}
\langle \kappa_v(m,T) \rangle = \left(\frac{\hbar M}{Z}\right) \Bigg\{ & \frac{(1-m)^2 c_s^3 \sum_k k \int \varphi(v)dv \partial \langle n_{1k} \rangle / \partial T}{\gamma_{1k}(v)} \\
& + m(1-m)\left(\frac{c^2}{T^2}\right) \frac{c_s \sum_k \sum_{k'} \int \varphi(v)kk' \langle n_{1k'} \rangle (1+\langle n_{1k'} \rangle)\gamma_{kk'}^* dv}{\gamma_{12k'}^{*2}(v)} \\
& + m(1-m)\left(\frac{c}{T^2}\right) \frac{c_s^2 \sum_k \sum_{k'} \int \varphi(v)kk' \langle n_{2k'} \rangle (1+\langle n_{2k'} \rangle)\gamma_{kk'}^{(21)} dv}{\gamma_{21k'}^2(v)} \\
& + m^2 c^3 \sum_k \int k \left(\frac{\partial \langle n_{2k} \rangle}{\partial T}\right) \frac{\varphi(v)dv}{\gamma_{2k}^{(ef)}} \Bigg\},
\end{aligned}
\tag{5.88}
$$

where the normalization factor is

$$
Z = \int \varphi(v)dv.
$$

Let the function $\varphi(v)$ have the form of uniform distribution, i.e.

$$
\varphi(v) = 1.
\tag{5.89}
$$

According to (5.89) and explicit equations for the attenuation coefficients (see formulae (5.56), (5.62), (5.66), (5.67), (5.68), (5.82) and (5.85)) we find

$$\langle \kappa_v(m,T) \rangle = R_1 + R_2 + R_3 + R_4\,, \tag{5.90}$$

where the dependences are

$$R_1 = (1-m)^2 \tag{5.91}$$

$$\times \left\{ \left(\frac{60\pi}{\gamma_1^2}\right) (l_f^2 c_d)^{-1} \left(\frac{Mc_s^2}{\hbar}\right) \varepsilon_1^{-2} \left(\frac{Mc_s^2}{T}\right) [(\nu_{f\,\max} - \nu_{f\,\min})] \right\},$$

$$R_2 = m(1-m) \left\{ 2.9 \times 10^{-4} \left(\frac{\varepsilon_1 \gamma_1}{\gamma_3}\right)^{-2} \left(\frac{\langle \theta_D \rangle}{Mc_s^2}\right)^2 \left(\frac{T}{Mc^2}\right)^2 \left(\frac{T}{\hbar}\right) \right.$$

$$\left. \times \Delta^8 c_d^{-1} J(T)(\nu_{f\,\max} - \nu_{f\,\min}) \right\}, \tag{5.92}$$

where

$$\varepsilon_1 = \frac{(r_0 - \langle r \rangle)}{r_0}\,, \quad \Delta = \frac{1}{l_{f\,\min}}\,, \tag{5.93}$$

and the function $J(T)$ is given by

$$J(T) = \int\limits_0^\infty \left[\frac{x^4}{(e^x - 1)}\right] \left[\frac{\partial^2}{\partial z^2} + \frac{(z/3)\partial^3}{\partial z^3}\right] \left[z^{-1/2} \mathrm{arctg}\left(\frac{2Tx}{\hbar c z^{1/2}}\right)\right] \mathrm{d}x\,. \tag{5.94}$$

The argument

$$z = \Delta^2 \quad \text{and} \quad M = \rho v_0\,, \tag{5.95}$$

where v_0 is the volume of the "elementary cell" ($v_0 \cong 10^{-20} - 10^{-22}$ cm^3).

$$R_3 = m(1-m) K_1^* \left(\frac{Mc_s^2}{T}\right) \frac{c_s n_0 l_{f\,\min}^5 k_D^4}{\gamma_3^2 N_f}\,, \tag{5.96}$$

where the parameters are

$$k_D = \frac{\langle \theta_D \rangle}{\hbar c_s}\,, \quad N_f = \frac{V}{v_f}\,,$$

$$K_1^* = 2 \times 10^{-3} \left\{ \frac{3}{2\pi} (\varepsilon^{1/2} - 1)^2 \right\}.$$

$$R_4 = m^2 \left(\frac{\pi^2}{6}\right) \left(\frac{T}{\hbar c}\right)^3 \frac{cL}{(1 + D_1 + D_2)}\,, \tag{5.97}$$

where the coefficients are

$$D_1 = 0.1\gamma_2^2 \left(\frac{LT}{Pd^{*2}}\right) \left(\frac{T}{\hbar c}\right)^6\,,$$

$$D_2 = \left[\frac{2(\varepsilon^{1/2} - 1)^2}{3\pi)} \left(\frac{L}{n_0}\right) \left(\frac{T}{\hbar c}\right)^4\right],$$

$$n_0 = \frac{N_0}{v_P}.$$

We shall underline that the formulae obtained are valid only at high temperatures when $T > \langle \theta_D \rangle$. The theoretical dependence $\langle \kappa(m, T) \rangle = \kappa(m, T)$ is illustrated in Fig. 5.4 for three possible cases.

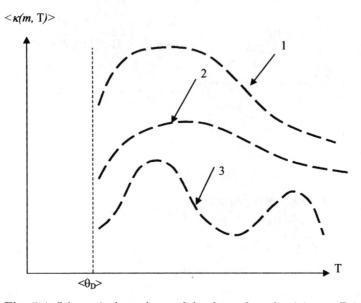

Fig. 5.4. Schematic dependence of the thermal conductivity coefficient of a porous substance in three physically different cases

Note that due to the nonlocality of the integral Eq. (5.12) it is reasonable to estimate the corresponding dimensions of the areas of this nonlocality of the mentioned formula for the relevant calculations. Besides, it is interesting to find boundaries of the transition of the relation (5.12) into the results which follow from the kinetic (local) theory of relaxation. We shall discuss only the momentum space. As is seen from Fig. 5.5 the operators of the thermal streams q_1 and q_2 may be considered local if their width Δk does not exceed the characteristic dimension of the area of intersection of these streams δk.

That is

$$\delta k \leq l^{-1} \leq \Delta k. \tag{5.98}$$

The condition of nonlocality, therefore, is the inverse

$$\delta k \gg \Delta k. \tag{5.99}$$

For the time τ over which the integration in (5.12) is done the inequality

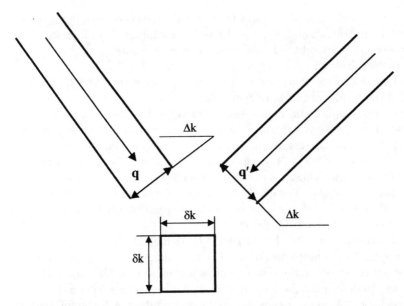

Fig. 5.5. Geometry of the incident thermal streams. Δk is the thermal stream width; Δk is the dimension of the local region in momentum space.

$$\delta\tau \gg \tau_{\text{rel}} \tag{5.100}$$

obviously must be valid. It is clear that only if the condition (5.100) is valid can we use the introduction of the local temperature T in the given fiber.

To conclude this section we emphasize the following:

1. the above procedure of calculation of $\kappa(m,T)$ allows us to find the coefficients of thermal conductivity for any highly disordered porous structures;
2. $\kappa(m,T)$ can qualitatively vary to a great extent depending on the relation between the involved parameters (see Fig. 5.4);
3. strong nonlocality in k-space is caused by the correlated interaction of the pore system and the main matrix.

5.4 Computer Analysis of the Formula for $\kappa(m,T)$

We derived the functional dependence of the thermal conductivity coefficient given in Fig. 5.4 only by the qualitative analysis of the very awkward general expression described by the formula (5.90) with fixed values of the parameters γ_1, γ_2 and γ_3. Plotting of the closest-to-reality curves for different values of these parameters may be made only by programming of the dependence $\kappa(m,T)$ and input of this program into the computer (for the program see Appendix A). It is quite clear that due to the "awkwardness" of the formula

(5.90) it will be the way out from the situation. The corresponding program was written in Fortran and then input into the computer. For different values of porosity m, temperature T and interaction coefficients γ_1, γ_2, and γ_3 the graph plotter drew the series of curves shown in Figs. 5.6–5.21.

And, at last, some words should be said concerning the physical interpretation of the results obtained in Sect. 5.3.

As every complex and highly non-ordered (in the wide meaning of the word) structure, porous structure is not an exclusion and here we should keep in mind the following. Imagine a comparatively simple situation (we already touched upon it in Sect. 4.7). Suppose there is a crystal dielectric with only one single spherical (chosen purely for convenience) pore of radius R, inside which there is some gas. The question is: how will its linear size influence such a physical parameter of a substance as the thermal conductivity coefficient of our strongly "discharged" dielectric?

Qualitatively it is quite obvious that if R is small, all the basic equilibrium and non-equilibrium properties of the structure almost do not change. Indeed, suppose we are studying the heat conduction of such a, by the way, quite real structure. To determine the temperature dependence of the thermal conductivity coefficient we need, as we already know from Sect. 4.7, the full pattern of microscopic internal relaxation. Here there are only two non-equilibrium systems, they are phonons and photons (for the sake of the purely qualitative pattern we ignore the reality and do not divide the phonons into longitudinal, transverse and optical). The phonons as oscillations of lattice sites provide the connection with the thermostat, whose role is traditionally played by the external environment at constant temperature T_0. Due to the interaction of the volume phonons with the boundary ones, the former have temperature T_0. This process, however, is not responsible for the functional dependence of the thermal conductivity coefficient κ, because besides this mechanism, there is the whole number of other not less important interactions to which we should refer: (a) interaction between longitudinal and transverse phonons (for which the characteristic time of energy relaxation is τ_{llt}), (b) change-of-state processes (umklapp processes) and (c) interaction of phonons with photons in the vicinity of a pore with characteristic time τ^*.

It is this latter mechanism for large R, comparable with the linear dimension L of the dielectric (but, of course, smaller!) that will make a decisive contribution to κ.

Indeed, let us represent κ as the sum $\kappa = (1-m)^2 \kappa_{11} + m(1-m)\kappa_{12} + m(1-m)\kappa_{21} + m^2 \kappa_{22}$, which is connected with the additive property of the phonon and photon heat flows (see above). Addend κ_{11} characterizes the heat conduction of a pure dielectric, for which, however, the dependence should take into account the sound velocity from c_s, from porosity m, defined by the known relation $m = V_2/V$, where $V = L^3$. κ_{12} and κ_{21} are the thermal conductivity of the boundary region, and κ_{22} is the thermal conductivity coefficient of the pore itself.

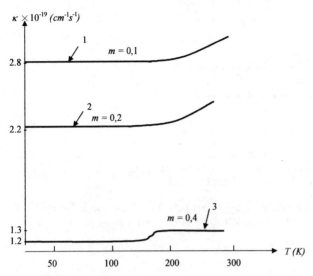

Fig. 5.6. Calculated curves of coefficient κ as a function of temperature for the following values of the parameters: $\gamma_1 = 1$, $\gamma_2 = 0$, $\gamma_3 = 100$. Curve 1 corresponds to porosity $m = 0.1$; curve 2 is $m = 0.2$; curve 3 is $m = 0.4$

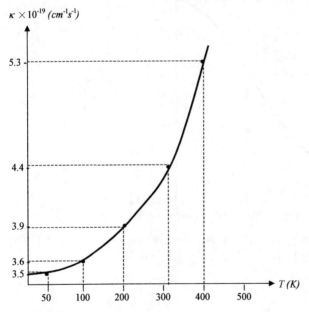

Fig. 5.7. Calculation curve of temperature dependence of thermal conductivity coefficient for $m = 0.9$, $\gamma_1 = 1$, $\gamma_2 = 0$, $\gamma_3 = 100$

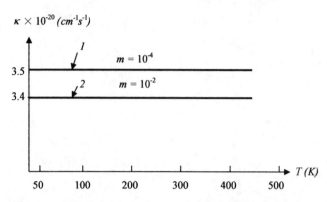

Fig. 5.8. Calculated lines for small values of porosity. Line 1 corresponds to porosity $m = 10^{-4}$, line 2 is $m = 10^{-2}$. In both cases the constants are $\gamma_1 = 1$, $\gamma_2 = 0$, $\gamma_3 = 100$.

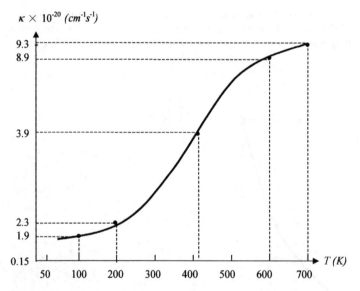

Fig. 5.9. Calculated curve for the limiting value of porosity $m = 0.999$. The constants are $\gamma_1 = 1$, $\gamma_2 = 0$, $\gamma_3 = 100$.

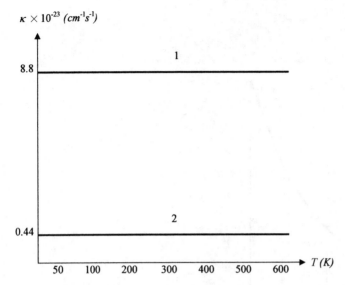

Fig. 5.10. Two calculated curves for $\gamma_1 = 0.1$, $\gamma_2 = 0$, $\gamma_3 = 10^6$

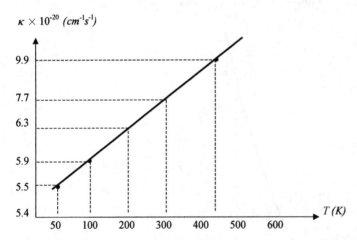

Fig. 5.11. Calculated dependence of the thermal conductivity coefficient for $m = 0.3$, $\gamma_1 = 10$, $\gamma_2 = 0$, $\gamma_3 = 100$

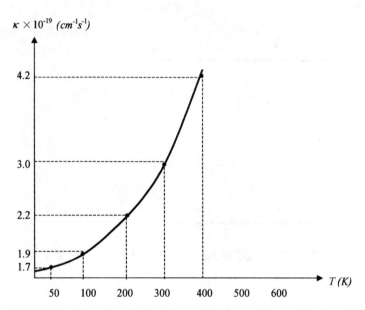

Fig. 5.12. Dependence of κ at $m = 0.3$, $\gamma_1 = 100$, $\gamma_2 = 0$, $\gamma_3 = 10^4$

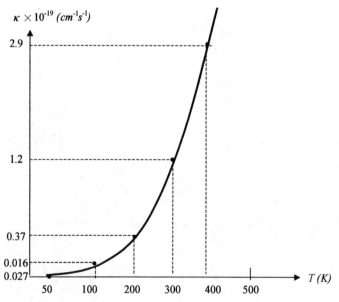

Fig. 5.13. Dependence of $\kappa\langle T\rangle$ for $m = 0.5$, $\gamma_1 = 10^{20}(\infty)$, $\gamma_2 = 0$, $\gamma_3 = 10^{40}(\infty)$

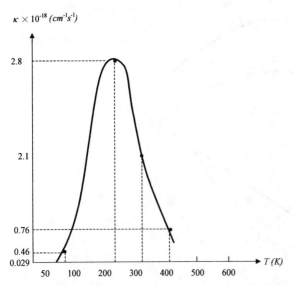

Fig. 5.14. Calculated behaviour of $\kappa\langle T\rangle$ for $m = 0.5$, $\gamma_1 = 10^{20}(\infty)$, $\gamma_2 = 10^{-4}$, $\gamma_3 = 10^{40}(\infty)$

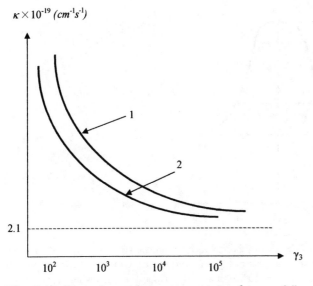

Fig. 5.15. Dependence of κ on parameter γ_3 for $m = 0.5$, $\gamma_1 = 100$, $\gamma_2 = 0$. Curve 1 corresponds to temperature $T = 1000\,\mathrm{K}$, and curve 2 corresponds to temperature $T = 300\,\mathrm{K}$.

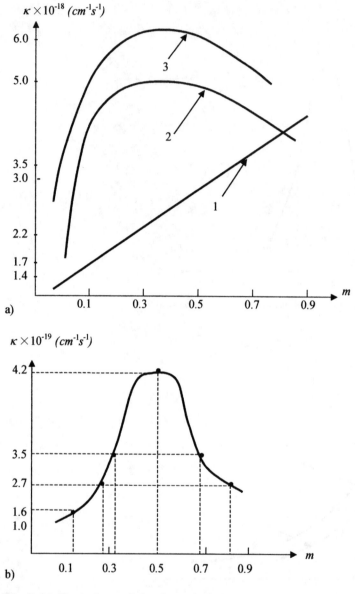

Fig. 5.16. Dependence of the thermal conductivity on the structure porosity m. (**a**) For all the three curves the parameters γ_1, γ_2 and γ_3 are equal and correspond to values $\gamma_1 = 10^3$, $\gamma_2 = 10^{-4}$, $\gamma_3 = 10^5$. Curve 1 describes the case when $T = 300\,\mathrm{K}$, curve 2 describes $T = 500\,\mathrm{K}$, curve 3 is $T = 700\,\mathrm{K}$ (**b**) Coefficient κ as a function of the parameter m for $T = 1000\,\mathrm{K}$, $\gamma_1 = 10^3$, $\gamma_2 = 10^{-4}$, $\gamma_3 = 10^5$.

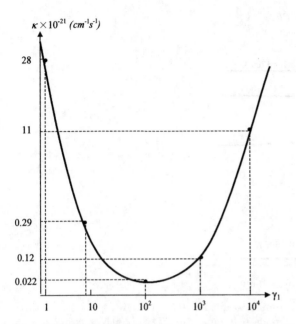

Fig. 5.17. Calculated dependence of κ on coefficient γ_1

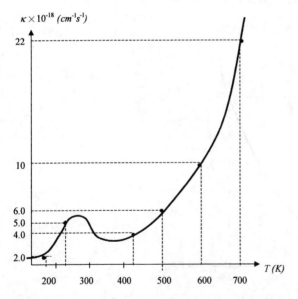

Fig. 5.18. Graphic dependence of the non-trivial behavior of κ for $m = 0.3$, $\gamma_1 = 100$, $\gamma_2 = 10^{-4}$, $\gamma_3 = 100$

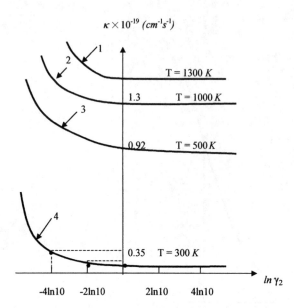

Fig. 5.19. Dependence of κ from $\ln \gamma_2$ at $m = 0.5$, $\gamma_1 = 10^3$, $\gamma_3 = 10^5$ in three different cases: curve 1 corresponds to temperature $T = 1300$ K, curve 2 is $T = 1000$ K, curve 3 is $T = 600$ K, curve 4 is $T = 300$ K

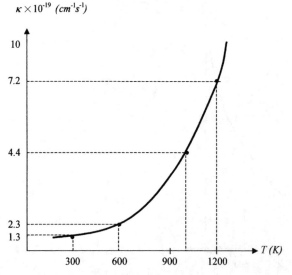

Fig. 5.20. Behavior of $\kappa\langle T\rangle$ for $m = 0.5$, $\gamma_1 = 10^3$, $\gamma_2 = 10^{-4}$, $\gamma_3 = 10^5$, $T = 300$ K

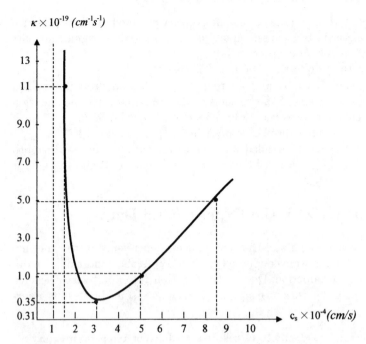

Fig. 5.21. Functional dependence of κ on the sound velocity in a porous substance for $m = 0.5$, $\gamma_1 = 10^3$, $\gamma_2 = 10^{-4}$, $\gamma_3 = 10^5$, $T = 300\,\mathrm{K}$

It is the region of the vicinity of the pore, according to the Bouguer–Lambert law, which is responsible for the absorption of non-equilibrium photons by a substance and is characterized by the coefficient h known from Sect. 4.7 and defined, by the way, as the product $c\tau_{21}$, where $\tau_{21} = \tau^*$. Since m is proportional to the pore volume, with the increase of Rm also increases and the contribution from the addend κ_{12} (or κ_{21}) becomes comparable with the contribution of the main matrix κ_{11}. It shows the necessity to take account of the region in the vicinity of the porous embedment indicating the a priori physical manifestation of non-equilibrium of black-body radiation.

Indeed, for equilibrium black-body radiation the Stefan-Boltzmann law is valid, according to which the energy is distributed according to the formula $E_2 = (4V_2/c)\sigma T_0^4$. As on the contact boundary the equilibrium is absent, we can write the following phenomenological equation: $\mathrm{d}E/\mathrm{d}t = (V_2/c)\sigma T^3 \mathrm{d}T/\mathrm{d}t = -\tau^{*-1}(E - E_2)$, where $\tau^* = h/c$.

The time τ^* was calculated above by the method of the non-equilibrium matrix density and the diagram technique using photon and phonon Green's functions. Here, the modernized Kubo formula (as applied to the case of a porous medium) was taken as the basis.

In conclusion it is worthwhile to draw the reader's attention once more to a very specific, from a methodological point of view, detail. The fact is that

Kubo's formula itself does not contain the porosity of a medium. The presence of the m-parameter in the general equation for κ is the most significant part in the present theoretical chapter.

And finally, the last thing to be noted is that the relaxation times themselves may be calculated by a more reliable, simple and properly verified method, that is by means of the method of the Boltzmann kinetic equation for quasi-particles. As far as the performance of the corresponding calculations is concerned, we have dwelled upon it in Chap. 4 (see Sect. 4.7).

The reader can find a detailed analysis of relaxation processes in both porous crystalline dielectrics and dielectrics in Appendices B and C.

5.5 Comparison with the Experimental Data

Experimental measurements of the temperature dependence of the thermal conductivity coefficient were conducted on eight types of cellulose of both domestic and foreign production. Despite the fact that the experimental scheme was well developed, the work on each sample consumed some hours plus labor-

Table 5.1. Thermal conductivity of cellulose at different temperatures (experiment)

N	Type of cellulose	Temperature T (°C)								
		50	75	100	125	150	175	200	225	250
1	Viscose sulphate low viscous coniferous (Baikal)	4.2	8.06	7.75	8.98	10.0	11.0	11.2	12.6	12.4
2	Viscose leaf-bearing BTIC (Bratsk timber-industrial complex)	9.95	12.5	10.7	9.97	11.5	11.8	13.4	14.4	15.6
3	Viscose sulphate bleached coniferous (Ust-Ilimsk)	7.9	8.95	8.34	8.81	9.96	11.3	–	–	–
4	E-2 MASHN	2.1	5.5	5.4	5.83	7.15	8.8	–	–	–
5	Sulphate coniferous 979 (Baikal)	18.4	14.5	13.9	14.7	–	–	–	–	–
6	Tyierdell (USA)	10.5	9.53	9.72	9.16	9.32	10.5	11.8	12.7	–
7	Viscose coniferous (Kotlas)	5.81	7.93	9.24	10.1	11.5	10.4	–	–	–
8	Finnish viscose (Raularapala)	12.4	12.4	14.4	12.8	13.7	13.9	14.0	16.8	–

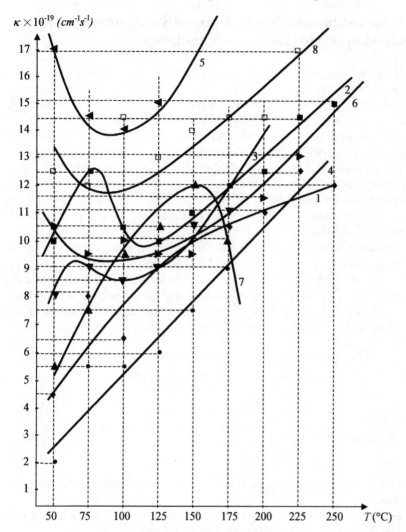

Fig. 5.22. Experimental curves obtained by measurement of the thermal conductivity of eight cellulose samples (Table 5.1)

consuming recomputation of the results with a calculator. After completion of all these calculations the final results of the measurements are shown in Table 5.1. The table allowed us to plot the experimental curves $\kappa_{\exp}(m, T)$ in the form of eight curves (see Fig. 5.22). Comparison of the obtained curves with the results of the computer processing of the theoretical formula show their satisfactory agreement both quantitative and qualitative. The latter statement may well be proved by direct comparison of curve 8 in Fig. 5.22 and Fig. 5.18.

5.6 Diffusion Saturation of Porous Structures by Liquid (Calculation of the Diffusion Coefficient)

In studying the process of saturation by liquid of various hydrophilic (water receptive) substances, of principle importance is the question of the calculation of the diffusion coefficient of such systems. Indeed, since the saturation velocity V_s is, according to Fick's law, $V_s = -D\nabla N$, where D is the diffusion coefficient and N is the number of molecules of a diffusing substance, then from the viewpoint of estimation of the absolute value of the saturation velocity it is necessary to make a preliminary microscopic analysis of the proportionality coefficient between both vectors. A simple estimation of D may be presented in such a way. Since in order of magnitude $V_s \cong D(N_1 - N_2)/\delta$, where δ is the thickness of the impregnated sample, N_1 is the number of liquid molecules coming at the initial moment of time on the surface, and N_2 is the number of molecules which reach the end, then knowing, for example, the average velocity V_s (the thickness δ is known, the time of percolation l_{per} is easy to fix by a stopwatch) we can define the "loss" of molecules ΔN as equal to the difference between the initial and final numbers of molecules $(N_1 - N_2)$ in the volume of the structure under investigation. This, in turn, allows us to make a forecast of the *absorption* capacity of the substance in question.

On the other hand, as the saturation time of the local area of porous matrix with thickness δ may be estimated by the formula

$$\tau_s = \frac{\delta^2}{D(m)}\,, \tag{5.101}$$

hence the porosity is

$$m = m(\tau_s, \delta, \{A\})\,, \tag{5.102}$$

where $\{A\}$ is the assemblage of all the other parameters including the microscopic ones.

Thus, calculating the coefficient $D(m)$, we shall be able to estimate from the saturation time τ_s the substance porosity m!

The difficulty of solution of this problem in a general form is, however, connected with the fact that the internal real structure of the fiber (fibril) matrix cannot be presented (at least to a first approximation) for all the possible structures in the form of some universal single model. And though such a hypothetical structure was suggested in Chap. 2 and then used further to explain a number of various experiments, the problem solved there had a somewhat different character. The thing is that in the impregnation of hydrophilic structures by a liquid the fiber begins to swell, thereby decreasing the volume of pores. That is why, on the one hand, there occurs the usual diffusion (though it may be connected with a capillary effect – everything depends on pore dimensions) at *fixed volumes* of *pores*, and on the other hand the purely hydrophilic process of *fiber swelling*. And if for the first process we may use diffusion terms for the second phenomenon, it is very problematic.

In order to find the solution taking into account both the filling of pores and their swelling we shall divide the problem into two stages. The first stage is to calculate the diffusion coefficient of a liquid by pores on condition that the pore walls are *absolutely solid*, and at the second stage we shall take into account *only swelling* of fibers, omitting the translator diffusion.

Stage 1. For absolutely solid walls we shall suggest as a working hypothesis a "fir-tree" model of pores (see [102, 103]) shown in Fig. 5.23. According to this figure the structure porosity is described in the form of some ensemble of randomly oriented cylindrical channels of different diameters branching, in turn, into thinner branches in all directions. The dimensions of the smallest channels are measured in micrometers and for them the diffusion consideration does nor suit because of a purely capillary effect. For this reason we shall introduce a limitation of the diffusion process by the formally introduced cri-

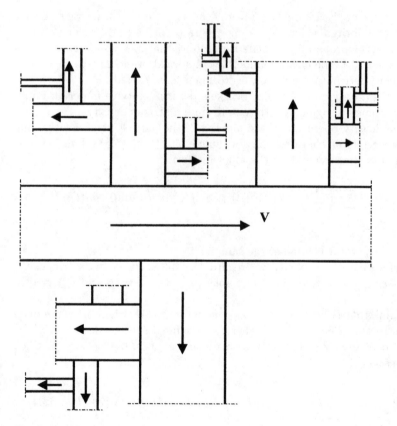

Fig. 5.23. Schematic pattern of a porous medium

tical radius of these microscopic pores – R_{cr}. Thus, at $R \leq R_{cr}$ there is no diffusion and at $R > R_{cr}$ it occurs.

In general we may state that in the absence of external pressure (the filtration connected with Darcy's law will be discussed in Sect. 5.8) the structure absorbes the liquid due to diffusion and the capillary effect (for more details see [11]).

In order to calculate $D(m)$ let us use the same approach as in Sect. 5.1, and again apply Kubo's formula for weakly non-equilibrium processes [76]. Within the framework of our current problem it will have the form

$$
D_{\alpha\beta} = \sum_{i=1}^{Q} \left(\frac{N_i}{Z_i}\right) \int_0^\infty vW(v)f(v)dv \int_v \int_0^\infty \langle [\delta i_\alpha(0,0), \delta i_\beta(\boldsymbol{x}, t)] \rangle d^3 x dt \,,
$$

$$(5.103)$$

where the operator of the deviation of the molecule stream from the equilibrium value is

$$
\delta i_{\beta\alpha}(\boldsymbol{x}, t) = \delta n(\boldsymbol{x}, t) V_\alpha(t) \,,
\tag{5.104}
$$

where $\delta n = n - n_{eq}$ is the local operator of the deviation of the particle density from the equilibrium value, V_α is the operator of molecule velocity ($\alpha = x, y, z$), W is the probability of penetration of the saturating liquid molecules into a branch, Q is the entire number of cylindrical branches (see Fig. 5.23), $f(v)$ is their distribution function over dimensions, v, and $Z_i = \int f(v)dv$ is the normalization factor. Angular brackets, as in the above material, imply thermodynamic averaging over the equilibrium matrix of density $\rho_0 = e^{-H/T}$.

As the density operators of the number of the particles and velocity due to their uncorrelated character may be "uncoupled", Eq. (5.6) taking account of (5.104) may be rewritten in the following form:

$$
D_{\alpha\beta} = V^{-1} \sum_{i=1}^{Q} \left(\frac{N_i}{Z_i}\right) \int_0^\infty \langle V^2(0) \rangle W(v) v^2 f(v)dv \int_0^\infty \langle [\delta n(0), \delta n(t)] \rangle dt \,,
$$

$$(5.105)$$

where V is the total volume of the sample.

For the sake of simplicity we regard the diffusion to be isotropic, i.e. $D_{\alpha\beta} = D\delta_{\alpha\beta}$, where Kroneker's symbol $\delta_{\alpha\beta} = 1$ at $\alpha = \beta$ and $\delta_{\alpha\beta} = 0$ at $\alpha \neq \beta$.

We think that $\langle V^2(0) \rangle$ may be taken equal to $3T/M$, where M is the mass of the molecules. The problem is therefore confined to the calculation of the internal correlator. We shall take as a postulate (it is, by the way, justified in most cases) that

$$
\langle [\delta n(0), \delta n(t)] \rangle = \left\langle [\delta n(0), \exp\{-i\int^t H_{\text{int}}(t)dt\}\delta n(0) \exp\{i\int^t H_{\text{int}}(t)dt\}] \right\rangle
$$

$$
= \exp(-\gamma t)\langle \delta n^2 \rangle \,,
\tag{5.106}
$$

where γ is the probability of the "death" of a molecule per unit time in its collision with the pore walls. For fluctuations (cf. [11], p. 373) we have

$$\langle \delta n^2 \rangle = -T \left(\frac{N_1^2}{v^2} \right) \left(\frac{\partial v}{\partial P} \right)_T . \tag{5.107}$$

Or, if we introduce the isothermal compressibility $\beta_T = -v^{-1}(\partial v/\partial P)_T$, we find

$$\langle \delta n^2 \rangle = \frac{T N_1^2 \beta_T}{v} . \tag{5.108}$$

Further, since W defines the probability of whether the liquid will go along the corresponding branch or not, it may be represented as the ordinary probability of overcoming an abstract potential barrier. Let us specify it phenomenologically by

$$W = \begin{cases} 1 & \text{at} \quad R > R_{cr} \\ 0 & \text{at} \leq R > R_{cr} . \end{cases} \tag{5.109}$$

From formulae (5.106)–(5.109) and taking account of (5.6) it follows that

$$D = \left(\frac{3T N_1^2 \beta_T}{\gamma M V} \right) \sum_{i=1}^{Q} \frac{N_i}{Z_i} \int\limits_{v_i > v_{cr}}^{\infty} v f(v) dv . \tag{5.110}$$

Let the law of pore distribution over volume be exponential, i.e.

$$f(v) = \exp \left(-\frac{v}{\langle v_i \rangle} \right) .$$

In this case

$$D = \left(\frac{3T N_1^2 \beta_T}{\gamma M V} \right) \sum_{i=1}^{Q} N_i \langle v_i \rangle \left(1 + \frac{v_{cr}}{\langle v_i \rangle} \right) \exp \left(-\frac{v_{cr}}{\langle v_i \rangle} \right) .$$

Or, introducing porosity m according to the relation

$$m = V^{-1} \sum_{i=1}^{Q} N_i \langle v_i \rangle \left(1 + \frac{v_{cr}}{\langle v_i \rangle} \right) \exp \left(-\frac{v_{cr}}{\langle v_i \rangle} \right) , \tag{5.111}$$

we obtain for the diffusion coefficient the simple formula:

$$D = \frac{3m T N_1^2 \beta_T}{\gamma M} . \tag{5.112}$$

It should be noted that in a number of cases the probability γ may depend on the momentum of the particles (molecules). Indeed, if we, for example, choose as the basic process the elastic scattering of the molecules by each other, it will appear that γ is proportional to the momentum $\boldsymbol{p}$ of the colliding molecule. If molecules interact according to the law of Casmir–Polder, $U(r) = -\alpha(r^2 + d_0^2)^3$, where d_0 is the "cutting" parameter, then $\gamma_p \Rightarrow p^{-3}$. We have assumed that the main mechanism of relaxation during the motion of a liquid about the "fir-tree" structure is related to a quasi-elastic mechanism of energy loss on the walls of *absolutely solid* pores. As a result of such numerous collisions, the molecules of the boundary layer of the liquid which are in direct contact with the walls diffuse along momentum (give their energy to the pores), and due to their interaction with the "volume" molecules, which are in direct contact with them, the "volume" molecules begin to defuse too.

Earlier we stated that the saturation time of such substances is defined by the dependence $\tau = L_{max}/V_s$, where L_{max} is the maximum distance covered by the liquid till it stops, and V_s is its average velocity. On the other hand, the saturation time defined by the diffusion mechanism is $\tau_s = L_{max}^2/D(m)$. Equating both relations and bearing in mind the gas-kinetic formula, $D = lv_T$, where l is the free path of molecules and v_T is their thermal velocity, we find a very important formula:

$$V_s = \frac{v_T l}{L_{max}}. \tag{5.113}$$

The given formula links to the parameters of the system: the saturation velocity (in other words the average velocity of liquid motion in the pores) which is a purely macroscopic parameter of the problem and the free path of the molecules. Therefore, by measuring the velocity V_s, knowing the length of the path L_{max} and the average thermal velocity v_T, we can estimate the average length of a molecular free path together with the scattering mechanism, for its probability is connected to l by the relation $w = v_T/l$.

From the viewpoint of experiment it would be interesting to measure the dependence v_s on the porosity m and also on the environment temperature. For we could, according to (5.112), by means of these known parameters estimate the isothermal compressibility of the liquid!

Stage 2. Let us now assume that the translation diffusion is absent and, instead, only the effect of pore swelling is present. How do we describe the swelling process purely mathematically? Let us again turn to the model and consider the simplified scheme of growing of the fiber wall thickness shown in Fig. 5.24. The hatched area corresponds to the swelling zone and the running radius (i.e. the radius of the decrease of free space defined by the radius R_1) is the formal parameter by which the swelling process will be described.

If r_0 is the largest radius of the swelling fiber, the equation characterizing the dynamical growth of the radius must have the form:

$$\frac{\partial r}{\partial t} = -\alpha(r - r_0), \tag{5.114}$$

where the velocity factor α corresponds to the velocity with which the liquid molecules add to the available chemical valent bonds (i.e. bound to the rate of chemical reaction K), localized on the pore surfaces. Thus, the minimum dimension r_0 is defined only by the chemical structure of a fiber.

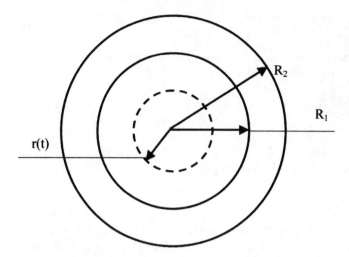

Fig. 5.24. Scheme of fibre swelling in the process of structure saturation by liquid

The solution of (5.14) with the initial condition $r(0) = R_1$ has the form

$$r(t) = r_0 + (R_1 - r_0)e^{-\alpha t} .$$

Since both the diffusion and the swelling processes have a probability character, the probability that both events will take place is determined by the product of the corresponding probabilities, and hence the effective probability of both events related to unit time is

$$W_{ef} = (\alpha\beta)^{1/2} . \tag{5.115}$$

In order to find α we shall reason in the following way. Let Q_1 be the number of unfilled active bonds, then the total number of active elements penetrating into the depth of a fiber over a distance of the order of the interatomic distance is

$$N_0 = N^* + Q_1 , \tag{5.116}$$

where $N^* = \langle a \rangle^2 l_f / \langle a \rangle^3 = l_f / \langle a \rangle$, where l_f is the fiber length and $\langle a \rangle$ is the average interatomic distance.

Let δr characterize the radial increase of a fiber as a result of its swelling. Therefore,

$$N_0 = \frac{\delta r l_f}{\langle a \rangle^2} ,$$

and, therefore, according to (5.116)

$$\frac{\delta r l_f}{\langle a \rangle^2} = Q_1 + \frac{l_f}{\langle a \rangle} ,$$

and hence

$$\delta r = \langle a \rangle + \frac{Q_1 \langle a \rangle^2}{l_f}. \tag{5.117}$$

On the other hand, because the product $K\delta r$ corresponds to the growth velocity due to the reactivity of a fiber, from (5.114) we get

$$\langle a \rangle K \left(1 + \frac{Q_1 \langle a \rangle}{l_f} \right) = -\alpha (r - r_0) = -\frac{\Delta r}{t_0},$$

where $t_0 = \alpha^{-1}$.

Taking, further, $\Delta r = r_0 - R_1$ and knowing that in the saturation region there are N_f fibers we find the time sought for:

$$t_0 = \frac{N_f (R_1 - r_0)}{\langle a \rangle K (1 + Q_1 \langle a \rangle / l_f)}. \tag{5.118}$$

The last step is to estimate the value of the radius r_0. As r_0 is the very critical dimension at which saturation as such stops (all Q_1 bonds are occupied), we may write

$$\Delta r = \frac{\delta r (R_2 - R_1)}{\langle a \rangle} = R_1 - r_0,$$

where R_2 is the fiber external radius.

Hence, using (5.117) we get

$$r_0 = R_1 - \left(1 + \frac{Q_1 \langle a \rangle}{l_f} \right) (R_2 - R_1). \tag{5.119}$$

Substituting (5.119) into (5.118) we find the time t_0 of interest to us as

$$t_0 = \frac{N_f (R_2 - R_1)}{\langle a \rangle K}. \tag{5.120}$$

We must emphasize that the obtained relation will be valid if all the fibers are similar on average. If the fibers are different, it is necessary to introduce the function of their dimensional distribution and to carry out the corresponding averaging. In this case the formula (5.120) is modified and it must be replaced by the following formula:

$$t_0 = \left[\frac{2N_f}{\langle a \rangle K (R_{\max} + R_{\min})} \right] \int_{R_{\min}}^{R_{\max}} f(R) R \, dR. \tag{5.121}$$

Substituting now (5.120) into (5.115), and the latter in turn into (5.112), we get the effective coefficient of diffusion, which accounts for not only pure diffusion but also pore swelling:

$$D_{ef} = \left(\frac{3N_1^2 T m \beta_T}{M} \right) \left[\frac{N_f (R_2 - R_1)}{\langle a \rangle K \gamma} \right]^{1/2}. \tag{5.122}$$

Application of this formula in practice requires knowledge of at least two parameters: the structure porosity m and the value γ. The other parameters may be taken as assigned. Indeed, the environment temperature is known, the liquid molecule mass is also known, the number of fibers and their dimensions may be determined by means of electron microscopy, and the liquid compressibility is also known. The velocity of adding (chemical reaction rate) K is considered as assigned (it may be found in reference books on chemistry).

If we are interested in the inverse problem, namely, if we want to estimate γ, we should do it in the following way. As $\tau_s = l/V_s = L^2/D$, taking into account (5.122) we find

$$\gamma = \frac{3N_1^4 T^2 m^2 \beta_T^2 N_f \delta R}{M^2 L^2 V_s^2 \langle a \rangle K}, \tag{5.123}$$

where $\delta R = R_2 - R_1$.

5.7 Determination of the Substance Porosity Coefficient

First of all we should clear up what is implied by the concept of the "self-consistent" determination of a substance porosity. As a rule, the "self-consistent" determination of some value means the non-contradictory introduction of some parameter (in our case m), when the various physical characteristics of a sample calculated using this parameter do not contradict either the experiment or the common sense.

As we already know, m is contained in the most important physical-technical characteristics of a particular substance. In order to collect all these magnitudes together we shall once more remind ourselves of the main physical properties which depend on the porosity:

1. sound velocity $c_s(m)$;
2. thermal capacity $c_p(m)$;
3. electric breakdown field $^{**}E(m)$;
4. ignition time delay $\tau_\iota(m)$;
5. thermal conductivity coefficient $\langle \kappa_v(m, T) \rangle$;
6. diffusion coefficient $D(m)$.

As an example we shall consider the set of the following physical parameters: the electric breakdown field $E^{**}(m)$, the thermal conductivity coefficient $\kappa(m, T)$ and the diffusion coefficient $D(m)$.

Suppose now that we have chosen some substance as the experimental sample, hereinafter referred to as substance "A". Now we want to measure all three chosen physical characteristics of the substance "A". For example, we shall first perform a statistical series of tests of the electric breakdown field $E^{**(\exp)}(m)$ at the specified ambient temperature T. Further, at the same fixed temperature we find the experimental value of the thermal conductivity $x^{(\exp)}(m)$. After this, using the values of $\kappa^{(\exp)}(m)$ and $E^{(\exp)}(m)$ we calculate by means of theoretical formulae (4.35) and (5.90) the value of the porosity m. At last, we calculate the diffusion coefficient $D^{(\exp)}(m)$ and also compare it with the theoretical value $D(m)$ given by the formula (5.122), into which we shall have to substitute the value of the porosity *found* from the two previous experiments. If as a result we shall see that there is adequate agreement between the values $D^{\text{theor}}(m)$ and $D^{(\text{exp})}(m)$, the conclusion will be obvious: the suggested method of determination of the structure porosity is effective and very reliable, and thus the *self-consistent* "void" index of the structure will characterize the *real* extent of filling of the substance "A" by fiber.

The described procedure for the determination of the average porosity may be carried over to any other physical characteristics of a chosen substance, and hence the correctness of any hypotheses which were involved to get formulae derived earlier may be verified. This is very important for proving the validity firstly of the model of f-points used to check particular calculations, and secondly for verification of the "viability" of a simplified model representation of such composite substances.

5.8 Darcy's Law (Microscopic Derivation)

In recent years investigations in the sphere of membrane technology have become very popular due to their applicability in studying various chemical and biochemical processes in biology, medicine, geophysics and engineering (see monographs [104–107] and papers [108–111]). As a rule, the physical laws used in the analysis of filtration processes are empirical. The best known among them is Darcy's law, binding the velocity of filtration V_F and the external pressure (∇P). Experimentally verified many times, this law is considered only phenomenological (at least we have no information about available publications devoted to its rigorous theoretical proof). But the modified Darcy's law may describe the nonlinear dependence of the velocity V_F on the pressure gradient. The coefficient of proportionality between V_F and ∇P is called the permeability coefficient and is denoted by K. K is a function of various parameters of the chemical mixture, such as the density ρ and the dynamical viscosity η.

It should be noted that the microscopic theory of filtration allows us to answer two principally important questions: (a) what is the real mechanism of internal relaxation responsible for variation of the filtration velocity when the liquid being filtered moves through a porous structure; and (b) what is temperature dependence of the permeability coefficient?

Mathematically, Darcy's law is

$$V = -K\nabla P, \tag{5.124}$$

where $K = A/\eta$, and the value A is defined by a particular structure of pores.

Thus, for example, if the porous medium is given as closely packed balls of diameter d, the constant A is determined by Slichter's formula, $A_s = d^2 m^2/96(1-m)$, or Coseni's formula, $A_c = d^2 m^3/96(1-m)^3$, where m is the porosity. If the filtering medium is approximated by close packing of thin cylinders, A is described by Purcell's formula, $A_p = d^2/32$.

It should be noted that when we deal with molecular dimensions, the geometry of the pores is not important because the free path of the molecules is small compared to the characteristic dimensions $d(l_M \ll d)$. In order to obtain the Eq. (5.124) strictly mathematically, we shall write the Boltzmann kinetic equation for the distribution function of molecules $f(\boldsymbol{p}, \boldsymbol{x}, t)$ in the following form:

$$\frac{\partial f}{\partial t} + \boldsymbol{v}\frac{\partial f}{\partial \boldsymbol{x}} + \boldsymbol{F}\frac{\partial f}{\partial \boldsymbol{p}} = L_1\{f\} + L_2\{f\}\,, \tag{5.125}$$

where $\boldsymbol{F}$ is the force acting on the molecules. We shall further hold that the external force is absent and $\boldsymbol{F} = 0$. The integral of binary molecule collisions is

$$L_1\{f\} = \int |\boldsymbol{v}_1 - \boldsymbol{v}_2|\sigma(\boldsymbol{v}_1, \boldsymbol{v}_2)(f_{1'}f_{2'} - f_1 f_2)\mathrm{d}^3 p_{2'}\,, \tag{5.126}$$

on condition that the laws of observation of energy and momentum are valid, i.e.

$$\begin{cases} \varepsilon(p_1) + \varepsilon(p_2) = \varepsilon(p_1') + \varepsilon(p_2') \\ \quad \boldsymbol{p}_1 + \boldsymbol{p}_2 = \boldsymbol{p}_1' + \boldsymbol{p}_2'\,, \end{cases}$$

where $\boldsymbol{v}_1$ and $\boldsymbol{v}_2$ are the velocities of the colliding molecules, their energy $\varepsilon = p^2/2M$, p is the momentum, M is its mass, $\sigma(\boldsymbol{v}_1, \boldsymbol{v}_2)$ is the collision cross-section.

The integral of collisions $L_2\{f\}$ takes into account the molecular collisions with the pore walls in which the stream being filtered moves with velocity $\boldsymbol{V}$. Its explicit expression is

$$L_2\{f\} = \Omega_p \int W(\boldsymbol{p}, \boldsymbol{p}')[f(p') - f(p)]\frac{\mathrm{d}^3 p'}{(2\pi\hbar)^3}\,, \tag{5.127}$$

at $\boldsymbol{p}' = \boldsymbol{p}\pm\boldsymbol{q}$, where Ω_p is the volume of one pore, $W(\boldsymbol{p}, \boldsymbol{p}')$ is the probability of quasi-elastic collisions of molecules with the pore walls per unit time (see also the above sections) and $\boldsymbol{q} = |\boldsymbol{p} - \boldsymbol{p}'|$ is a small change of molecular momentum as a result of collisions with the walls.

The velocity $\boldsymbol{v}$ used in (5.125) in the attendant system of coordinates characterizes the molecules' own velocity. In passing to the system of coordinates moving with the stream velocity $\boldsymbol{V}$ relative to a fixed observer, in the observer's frames of reference it should be held that $\boldsymbol{v} = \boldsymbol{v} + \boldsymbol{V}$. Taking into account the above statement the Eq. (5.125) can be written as

$$\frac{\partial f}{\partial t} + (\boldsymbol{v} + \boldsymbol{V})\frac{\partial f}{\partial \boldsymbol{x}} = L_1\{f\} + L_2\{f\}\,. \tag{5.128}$$

It should be noted that the addend of the form $\boldsymbol{V}\partial f/\partial \boldsymbol{x}$ in the derivation of the Navies–Stokes equation leads to the nonlinear term $(\boldsymbol{V}\nabla)\boldsymbol{V}$ characterizing the substantional derivative which is on the left-hand side. As we deal with macroscopic time $\delta t = L/V \gg \tau_1$, where τ_1 is the characteristic time of achievinga quasi-equilibrium state in the system of molecules and L is the average length of the pores, the solution of (5.128) should be sought in the form $f = \langle f \rangle + \delta f$, where

$$\langle f \rangle = \exp\left(-\frac{[\varepsilon(p) + \boldsymbol{pV}]}{T}\right). \tag{5.129}$$

The order of smallness of the magnitude $\delta f/f$ is l_2/L, where l_2 is the free path of the molecules at their collision with the pore walls (addend $L_2\{f\}$ on the right-hand side of (5.128)).

In order to relate the macroscopic force of friction, which appears in the process of viscous substance motion through the structure, to the change of the distribution function, which is due to the quasi-elastic collisions of molecules with pores, we should use the general equation for the dissipative function of the classic Boltzmann gas [11]:

$$T\frac{\mathrm{d}S}{\mathrm{d}t} = -T\int\left(\frac{\mathrm{d}f}{\mathrm{d}t}\right)\ln f\,\mathrm{d}^3p, \tag{5.130}$$

where S is the entropy.

Since $\mathrm{d}f/\mathrm{d}t = -\delta f/\tau_2$, where, by order of magnitude $\tau_2 = W^{-1}$, and $\delta f = f - \langle f \rangle$, taking

$$T\frac{\mathrm{d}S}{\mathrm{d}t} = \int_0^{\boldsymbol{V}} F_{fr}\mathrm{d}\boldsymbol{V}, \tag{5.131}$$

from (5.130) we find a very important formula relating the *macroscopic* force of friction to the change of the *distribution function* of the molecules:

$$\boldsymbol{F}_{fr} = T\Omega_p\left(\frac{\partial}{\partial\boldsymbol{V}}\right)\int_0^{\boldsymbol{V}}\delta f^2\frac{\mathrm{d}^3p}{\langle f\rangle\tau_2}. \tag{5.132}$$

Substituting the solution (5.129) into (5.128) and using the smallness of $\boldsymbol{V}$, we easily get for the stationary case when $\mathrm{d}f/\mathrm{d}t = 0$ that $\delta f = [\boldsymbol{v}\nabla(\boldsymbol{pV})]\tau_2 f_0/T$. If we take the z axis along the axis of a cylindrical pore, then the flux velocity will be $V_z(\boldsymbol{V} = (0, 0, V_z)$. The radial components of the molecular velocity (momentum) are $v_r(p_r)$ and hence

$$\delta f = v_r p_z \left(\frac{\partial V_z}{\partial r}\right)\left(\frac{\tau_2 f_0}{T}\right), \tag{5.133}$$

where the function

$$f_0 = \langle f\rangle|_{\boldsymbol{V}=0} = \mathrm{e}^{-\varepsilon(p)/T}.$$

Since at small dimensions we may consider that $\partial V_z/\partial r = -V_z/\mathrm{d}$, then using the solution (5.133) we get from (5.132) the following equation for the frictional force related to one pore:

$$F_{fr} = \frac{(2\pi\tau_2 V_z)}{(mM^2 T \mathrm{d}^2)} \int_0^\infty p^3 \mathrm{d}p \int_{-\infty}^\infty f_0^2 p_z^2 \frac{\mathrm{d}p_z}{\langle f \rangle} \,. \tag{5.134}$$

The simple integrals in Eq. (5.134) are readily calculated if we take into account f_0 and $\langle f \rangle$. Finally,

$$F_{fr} = \frac{2\nu_2 V_z M (1 + 4MV_z^2/3T) \exp\{MV_z^2/2T\}}{\mathrm{d}^2 m} \,, \tag{5.135}$$

where the kinematic viscosity is $\nu_2 = \tau_2 v_T^2$ and the middle heat velocity of the molecules is $v_T = (T/M)^{1/2}$. Note that in (5.135) we take into account that the equilibrium distribution function is $f_0 = (2\pi MT)^{-3/2} \exp\{-p^2/2MT\}$.

On the other hand, if we accept the condition $F_{fr}/\Omega = -\partial P/\partial z$, where P is the pressure, then from (5.135) we immediately obtain directly Darcy's non-linear law:

$$\varphi(V_z) = -\frac{(A/\nu_2)\partial P}{\partial z} \,, \tag{5.136}$$

where the functions

$$\varphi(V_z) = V_z \exp\left[\left(\frac{MV_z^2}{2T}\right)\left(1 + \frac{4MV_z^2}{3T}\right)\right] \,, \tag{5.136a}$$

and

$$A = \frac{\mathrm{d}^2 m\Omega}{2M} \,. \tag{5.136b}$$

It should be said that for the description of a real filtration process, in the function $\varphi(V_z)$ a formal substitution should be done. Namely, we should consider that $M \Rightarrow \mu = \rho\Omega_p(b/l_1^3)$, where ρ is the density of the substance being filtered, b is a molecule diameter and l_1 is the average free path for binary collisions. Then the function renormalized in such a way will be $\varphi(V_z) = V_z \exp\{\mu V_z^2/2T\}(1 + 4\mu V_z^2/3T)$, and at $T \gg \mu V_z^2$ we have

$$V_z \left(1 + \frac{11\mu V_z^2}{6T}\right) = \left(\frac{A}{\nu_2}\right)\left(\frac{\partial P}{\partial z}\right) \,. \tag{5.137}$$

The obtained formulae allow us to make two important conclusions:

1. The microscopic approach using the classic Boltzmann equation has allowed us to give a rigorous mathematical basis for the empirical Darcy's law and, moreover, to find theoretically nonlinear dependence of the filtration velocity on the external pressure gradient.
2. From (5.136) it follows that at ν_2, which changes depending on temperature according to the law $T^{-1/2}$, the filtration velocity V_z grows as $T^{1/2}$ (see Fig. 5.25). If ν_2 depends on T not as $T^{-1/2}$ but otherwise, then, naturally, the qualitative behavior of V_z will change.

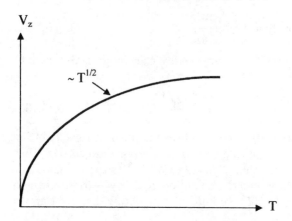

Fig. 5.25. Temperature dependence of the filtration velocity

5.9 Influence of the High-Frequency Deformation Wave on the Filtration Process

The approach outlined in Sect. 5.8. will enable us to find the dependence of the filtration velocity on the *frequency* of the acoustic disturbance affecting a two-dimensional porous film. It should be noted that in reality the acoustic influence can manifest itself as pulses of various forms. The deformation wave may be (a) periodic, i.e. regular, (b) irregular, e.g. a solution, and (c) an impulse wave, acting in some local area of a surface. We shall be concerned with a purely periodical wave, though the theory presented below allows us to account for external influences of *any* type.

According to Fig. 5.26 imagine the following picture. Let a periodic deformation wave propagate over the surface. The deviation of the specified element of the surface from the equilibrium form will be denoted as $\xi(x, y, t)$.

We shall also believe that the radius of a specified pore r depends on time according to the assigned law $r(t)$ – the pore "breathes". Therefore, Boltzmann's equation can be written as follows:

$$\frac{\partial f}{\partial t} + \left(v_z + V_z + \frac{d\xi}{dt}\right)\frac{\partial f}{\partial z} + \left(v + \frac{dr}{dt}\right)\frac{\partial f}{\partial r} = L_1\{f\} + L_2\{f\}. \quad (5.138)$$

In the τ-approximation the solution of (5.138) may be found easily. Indeed, according to (5.133) we have

$$\delta f = \left(\frac{\tau_2}{MT}\right)\left\{p_z\left(v_z + V_z + \frac{d\xi}{dt}\right)\left(\frac{\partial V_z}{\partial z}\right)\left(v + \frac{dr}{dt}\right)p_z\left(\frac{\partial V_z}{\partial r}\right)\right\}f_0. \quad (5.139)$$

If V_z does not depend on the z coordinate and $\partial V_z/\partial r = -V_z/d$, we find

$$F_{fr}\left\{\frac{dr}{dt}\right\} = \frac{2\Omega\nu_2 MV_z(1 + 11\mu V_z^2/6T)[1 + \mu(dr/dt)^2/T]}{d^2m}. \quad (5.140)$$

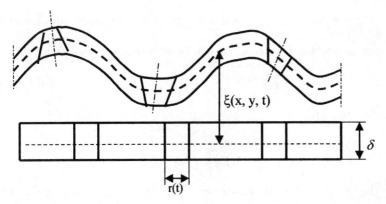

Fig. 5.26. Schematic pattern of a propagating deformation wave on the surface of a porous film (membrane)

It is interesting to note that if the longitudinal component of velocity V_z does not depend on z, the frictional force does not depend on the displacement ξ but depends only on the compression (expansion) of the pores by means of the function $r(t)$. The analysis of the dependence (5.140) shows, in particular, that at temperatures $T \gg \mu V_z^2$ (if $V_z = 10 \,\text{cm/s}$ and $\mu = 10^{-18} \,\text{g}$ then $\mu V_z^2 = 1°\,\text{K}$) the following formula may be found:

$$V_z = -\left[\frac{A}{\nu_2}\right]\left(\frac{\partial P}{\partial z}\right)\left[1+\mu\left(\frac{dr}{dt}\right)^2\right]^{-1}\Bigg|_{r=d/2} . \tag{5.141}$$

At very low temperatures $(T \ll \mu V_z^2)$ the relation (5.10) including the equation $F_{fr} = -\Omega \partial P/\partial z$ will be reduced to a transcendental equation whose solution describes the dependence of 5.133percolation velocity on the pressure gradient.

Let us analyze the formula (5.141). We have two cases as follows.

1. Let the frequency of the periodic external oscillation satisfy the condition $\omega \gg V_z/\delta$, where δ is the film thickness.
    At these frequencies the relation (5.141) should be time averaged. It means that the average value of the filtration velocity should be defined as

$$\langle V_z \rangle = (2T_0)^{-1} \int_{-T_0}^{T_0} V_z dt \,,$$

where the period $T_0 = 2\pi/\omega$.
Then we may write

$$V_z = -\left[\frac{A_0}{\nu_2}\right]\left(\frac{\partial P}{\partial z}\right)\left\langle d^2(t)\left[1+\frac{\mu\omega^2 d^2(t)}{4T}\right]^{-1}\right\rangle , \tag{5.142}$$

where the new function A_0 does not depend on the pore radius and $A_0 = A/d^2(t)$.

Let, for example, $d(t) = d_0 \cos \omega t$. Then by elementary averaging we find

$$V_z(\omega) = -\left[\frac{A_0}{\nu_2}\right]\left(\frac{\partial P}{\partial z}\right) J(\omega, T), \tag{5.143}$$

where the function

$$J(\omega, T) = 2\left[\frac{(q+1)}{\pi q}\right]\left[\int_0^{\pi} \frac{du}{(1 + q \sin^2 u)} - \frac{2}{q}\right],$$

where the dimensionless parameter $q = \mu \omega^2 d_0^2 / 4T$. By substitution of $x = tg(u/2)$ this integral can be rewritten as

$$J(\omega, T) = 4\left[\frac{(q+1)}{\pi q}\right]\int_{-\infty}^{\infty} \frac{(1 + x^2)dx}{(1 + x^2)^2 + 4qx^2} - \frac{2}{q}. \tag{5.144}$$

The denominator has two complex poles in the up plane of the complex variable z, which are $x_{1,2} = i[(q+1)^{1/2} \pm q^{1/2}]$. The theory of deduction give us the following expression: $J(\omega, T) = [2(q+1)^{1/2}/q] - 2/q$. The asymptotic forms of this function in the two limiting cases are the following:

$$J(\omega, T) = 2\begin{cases} 1 - \dfrac{q}{4} \text{ at } q \ll 1 \left(\dfrac{V_z}{\delta} \ll \omega \ll \left(\dfrac{2}{d_0}\right)\left(\dfrac{T}{\mu}\right)^{1/2}\right), \\[3mm] \dfrac{1}{q^{1/2}} - \dfrac{1}{q} \text{ at } q \gg 1 \left(\omega \gg \left(\dfrac{2}{d_0}\right)\left(\dfrac{T}{\mu}\right)^{1/2}\right). \end{cases} \tag{5.145}$$

The obtained result is notable because it allows us to make a very important practical conclusion. At large frequencies when

$$\omega \gg \left(\frac{2}{d_0}\right)\left(\frac{T}{\mu}\right)^{1/2} \tag{5.146}$$

which for parameters $T = 300\,\mathrm{K}$, $\mu = 10^{-18}\,\mathrm{g}$, $d_0 = 10^{-3}\,\mathrm{cm}$ corresponds to the range $\omega \gg 10^5\,\mathrm{Hz}$, the filtration practically *stops*.

2. Now let frequency $\omega \ll V_z/\delta$.

In this case averaging cannot be done because the characteristic filtration time is much less than the period of the deformation wave. The latter means that the percolation occurs faster than the deformation wave may propagate along the surface and, therefore, a porous film will have no time even to "feel" external action. Indeed, inserting in the formula (5.141) $\omega = 0$ $(d = d_0)$, we get the relation (5.136).

Putting it exactly, the above formulae will be valid if all the pores are similar. It is clear that such an ideal case usually is not encountered and the dimension spread of the pores should be taken into account. If we introduce a distribution function $\varphi(d)$, one more averaging should be done in the formula (5.142). The particular type of the dependence $\varphi(d)$ will not change the results (5.143)–(5.145) too much.

Sometimes the microscopic expression for the permeability coefficient K obtained by means of the method of the density matrix may appear helpful (see Sect. 5.1). Not to exhaust the reader we shall at once give the formula sought for:

$$(K^{-1})_{\alpha\beta} = T^{-1} \int \int \langle [\nabla_\alpha \varepsilon(0,0), \nabla_\beta \varepsilon(\boldsymbol{x},t)] \rangle \, \mathrm{d}^3 x \mathrm{d}t \,, \qquad (5.147)$$

where the operator of differentiation $\nabla_\alpha = \partial/\partial x_\alpha$, the square brackets mean the commutator, and $\varepsilon(\boldsymbol{x},t)$ is the operator of energy density written in Heisenberg's representation. Note, that the relation (5.147) determines not the coefficient itself, but the inverse matrix from it. After calculating the commutators and integration we should simply take $(K^{-1})_{\alpha\beta}$.

Summarizing, we note that

1. The microscopic theory of filtration based on the application of the classic kinetic Boltzmann equation allowed us to find firstly the temperature behavior of the average velocity $V_{Ft}(V_z)$, and secondly, to find nonlinear over V_z amendments to empirical Darcy's law which show themselves at relatively high velocities.
2. The dependence of the filtration velocity found as the frequency function of the deformation wave and schematically shown in Fig. 5.27 indicated

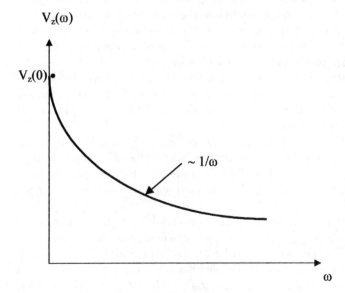

Fig. 5.27. Dependence of the average filtration velocity on the frequency of an external alternating acoustic disturbance

that at great frequencies the filtration process may stop completely. This is to be taken into account particularly in investigating, for example, biological objects put into high-frequency acoustic fields.

5.10 On the Connection of the Joule–Thomson Effect with Darcy's Law

It should be mentioned that in studying the diffusion saturation of porous structures by a liquid (see Sect. 5.6) the conditions which the so-called Joule–Thomson effect begins to work are quite real. Its essence is described practically in every textbook on statistical physics, which is why we shall very briefly review the basic physical content of this phenomenon.

Suppose that there is a cylinder completely heat-insulated and divided in an arbitrary place into two parts by porous (say, of compact paper) partition. Now assume, that in the left-hand part a pressure P_1 and temperature T_1 are created, and in the right-hand one P_2 and T_2, respectively. Let now a piston from the left begin to compress the liquid in the left-hand space V_1 (V_1 is its volume), and the liquid begins slowly to leak through the porous medium (which we shall sometimes call the membrane) into the right-hand part, increasing its volume V_2. In this picture the dependence immediately manifests itself of the pressure on the temperature on both sides of the porous partition. If we denote the temperature difference $T_1 - T_2$ as δT and the pressure difference $P_1 - P_2$ as δP, then the relation between them will be $\delta T = \alpha \delta P$. Here, due to a small change of temperature and pressure, the coefficient α formally may be presented as the coefficient of expansion in a Taylor series expansion of the function $T(P)$. And hence it is the partial derivative $(\partial T/\partial P)_X$, where X is an unknown variable remaining constant in the process under consideration and which we have to determine.

It is quite obvious that in this process the entropy is not conserved, because in reality there is always friction of liquid in the partition pores, though according to the problem conditions, the whole cylinder is heat-insulated and the global entropy is constant!!

Let us denote the internal energy from the left as E_1 and from the right as E_2. It is clear that the energy E_1 must be higher than E_2. The difference between the energies E_1 and E_2 (ΔE), according to the well-known virial theorem will be defined by the work connected with the change of thermodynamic potential.

Now the question is, what particular potential is most convenient for us in the given case. In general, it is quite obvious that in our variables, namely P, V and T, the most convenient is the potential $\Omega(T, V, \mu(P, T))$, where μ is the chemical potential. By definition we have $\Omega = -PV$, and hence $(\delta E)_{S,V,N} = (\delta \Omega)_{\mu,V,T}$ and $E_1 - E_2 = \Omega_1 - \Omega_2 = P_2 V_2 - P_1 V_1$. From this it directly follows that $E_1 + P_1 V_1 = E_2 + P_2 V_2 = \text{const}$. The sum $E + PV$, as we know from Sect. 2.4, is the enthalpy W, and therefore we conclude that in the Joule–Thomson process the enthalpy is conserved and the unknown variable $X = W$. Thus $\delta T = (\partial T/\partial P)_W \delta P$. In order to estimate the derivative $(\partial T/\partial P)_W$ we should pass over from the variables (P, W) to more convenient variables, for example (P, T). Then

$$\alpha = \alpha_W = \left(\frac{\partial T}{\partial P}\right)_W = \frac{\partial(T, W)}{\partial(P, W)}\frac{\partial(T, P)}{\partial(T, P)} = \frac{\partial(T, P)}{\partial(P, W)}\frac{\partial(T, W)}{\partial(T, P)}$$

$$= -\left(\frac{\partial W}{\partial P}\right)_T\left(\frac{\partial T}{\partial W}\right)_P .$$

Further, since $dW = TdS + VdP$, then $(\partial W/\partial T)_P = T(\partial S/\partial T)_P = C_P$ and $(\partial W/\partial P)_T = V + T(\partial S/\partial P)_T$.

In order to determine the derivative $(\partial S/\partial P)_T$, let us use the Gibbs thermodynamic potential. Its differential is $d\Phi = -SdT + VdP$. Therefore, $(\partial\Phi/\partial T)_P = -S$, $(\partial\Phi/\partial P)_T = V$. Equating the second mixed derivatives we find that $(\partial S/\partial P)_T = -(\partial V/\partial T)_P$.

Gathering all the data together we finally find for the Joule–Thomson coefficient the following formula:

$$\alpha_W = \left(\frac{\partial T}{\partial P}\right)_W = C_P^{-1}\left[T\left(\frac{\partial V}{\partial T}\right)_P - V\right]. \tag{5.148}$$

From this it is clearly seen that for the ideal gas from the equation of state $PV = NT$, it follows that $(\partial V/\partial T)_P = N/P = V/T$, and hence $\alpha_W = 0$. The interpretation of such a result is quite evident: since for an ideal gas there is no interaction with anything at all, including the pore surfaces, there is no temperature change. The temperature as well as the pressure both from the left and from the right of the membrane will be the same.

So, the brief description given above enables us to make a very important conclusion, that the Joule-Thomson coefficient α_W is a purely dissipative characteristic of an irreversible thermodynamic process, and therefore we can try to relate α_W to the permeability coefficient K used in Darcy's formula (5.124), which depends, as we know, on the viscosity η.

Let us find the relation between α_W and K strictly mathematically. For this purpose let us calculate the dissipative function dQ/dt. By definition we have $dQ/dt = TdS/dt$, where S is the entropy. For a classical Boltzmann's gas of molecules the general expression for the entropy through the non-equilibrium distribution function f_p has the form:

$$S = -\frac{\int f_p ln(f_p/e)d\Gamma}{\int \langle f_p\rangle d\Gamma}, \tag{5.149}$$

where the element of phase-space volume is $d\Gamma = Vd^3p$, and the equilibrium Boltzmann distribution function is

$$\langle f_p\rangle = \exp\left\{-\frac{\varepsilon_p}{T_0}\right\}, \tag{5.150}$$

where the energy of the translation motion of the molecules is $\varepsilon_p = p^2/2M$ and T_0 is the equilibrium temperature.

Firstly, it is seen that the volume V falls out from the answer, and secondly, the integral in the dominator is easily calculated and gives

$$Z = \int \langle f_p \rangle \mathrm{d}^3 p = \int_0^\pi \langle f_p \rangle p^2 \mathrm{d}p 4\pi = (2\pi M T)^{3/2} \,.$$

Differentiating (5.149) with respect to time, taking into account the above, we find

$$\dot{S} = - \int \dot{f}_p \ln f_p \frac{\mathrm{d}^3 p}{Z} \,. \tag{5.151}$$

According to the Boltzmann equation we have $\dot{f}_p = L\{f_p\}$, where $L\{f_p\}$ is the integral of classical gas molecule collisions. In the τ-approximation $L\{f_p\} = -\delta f_p/\tau_p$, where τ_p is the relaxation time and $\delta f_p = f_p - \langle f_p \rangle$.

Further, as the total derivative is

$$\dot{f}_p = \frac{\partial f_p}{\partial t} + v \left(\frac{\partial f_p}{\partial r} \right) + F \left(\frac{\partial f_p}{\partial p} \right) = - \frac{\delta f_p}{\tau_p} \,,$$

then taking the force F acting on the molecules to be equal to zero and taking the distribution to be stationary, for times $\delta t \gg \langle \tau_p \rangle$ we find for the sought-for correction δf_p that

$$\delta f_p = -\tau_p v \nabla f_p^* \,, \tag{5.152}$$

where the quasi-equilibrium distribution is introduced by the equation

$$f_p^* = \exp \left\{ -\frac{(\varepsilon_p - pV)}{T} \right\} \,, \tag{5.153}$$

where $V(r,t)$ and $T(r,t)$ are the non-uniform distributions of the macroscopic flow velocity and the stream temperature respectively.

From (5.152), therefore, in the linear approximation over small gradients we obtain the sought-for correction to the equilibrium distribution function:

$$\delta f_p = - \left(\frac{\tau_p \langle f_p \rangle}{T_0} \right) \left[(p\nabla)(vV) + \left(\frac{v\nabla T}{T_0} \right) (\varepsilon_p - pV) \right] \,. \tag{5.154}$$

Let us now return to the formula (5.150). Substituting in it $f_p = \langle f_p \rangle + \delta f_p$, and considering the total time derivative equal to $\mathrm{d}f_p/\mathrm{d}t = \mathrm{d}f_p^*/\mathrm{d}t = -\delta f_p/\tau_p$, and using the expansion of the natural logarithm into a Taylor series, according to which $\ln(1+x) \approx x$ for small x, for the dissipative function we find

$$\dot{Q} = T_0 \dot{S} = \left(\frac{T_0}{Z} \right) \int \left(\frac{\delta f_p}{\tau_p} \right) \ln(\langle f_p \rangle + \delta f_p) \mathrm{d}^3 p \approx J_1 + J_2 \,, \tag{5.155}$$

where

$$\begin{cases} J_1 = \left(\dfrac{T_0}{Z}\right) \displaystyle\int \left(\dfrac{\delta f_p}{\tau_p}\right) \ln\langle f_p\rangle \mathrm{d}^3 p \,, \\[4mm] J_2 = \left(\dfrac{T_0}{Z}\right) \displaystyle\int \left(\dfrac{\delta f_p^2}{\langle f_p\rangle \tau_p}\right) \mathrm{d}^3 p \,. \end{cases}$$

Let us consider each term separately, beginning with J_1.

So, by means of (5.154) and taking into account that the equilibrium distribution function is $\langle f_p\rangle = \exp\{-\varepsilon_p/T_0\}$, we have

$$J_1 = \left(\frac{T_0}{Z}\right) \int \left(\frac{\delta f_p}{\tau_p}\right) \ln\langle f_p\rangle \mathrm{d}^3 p$$

$$= -(ZT_0)^{-1} \int \langle f_p\rangle \varepsilon_p \left\{ (p\nabla)(vV) + (v\nabla T)\frac{\varepsilon_p}{T_0} \right\} \mathrm{d}^3 p \,.$$

Due to integration with respect to the angular variables the second term proportional to the temperature gradient disappears.

Using now the rule of averaging over angular variables, namely the condition that $\int v_i p_k(\ldots)\mathrm{d}^3 p = (\delta_{ik}/3) \int \boldsymbol{vp}(\ldots)\mathrm{d}^3 p$, the first term may be transformed to the form

$$J_1 = - \left(\frac{\mathrm{div}V}{3ZT_0}\right) \int pv\langle f_p\rangle \mathrm{d}^3 p \,. \tag{5.156}$$

If we now remember one of the basic equations of hydrodynamics, the equation of continuity, according to which $\partial\rho/\partial t + \mathrm{div}(\rho V) = 0$, where ρ is the liquid density and consider the liquid to be incompressible (i.e. Newtonian), it should be taken that $\rho = \mathrm{const}$ and we at once find $\mathrm{div}V = 0$. The latter condition just demonstrates that the integral J_1 turns to zero. So, $J_1 = 0$.

Returning now again to (5.155), we finally obtain

$$\overset{\bullet}{Q} = \int \delta f_p^2 \frac{\mathrm{d}^3 p}{\tau_p\langle f_p\rangle} \,,$$

which, as we see, with all the parameters is not a negative magnitude and does not contradict the second law of thermodynamics. Now we substitute here the solution (5.154). As a result we get

$$\overset{\bullet}{Q} = (ZT_0)^{-1} \int \tau_p\langle f_p\rangle \left[(v\nabla T)\frac{(\varepsilon_p - pV)}{T_0} + (p\nabla)(vV) \right]^2 \mathrm{d}^3 p \,.$$

Raising the expression in square brackets to the second power and leaving only even powers of products p and v (odd powers in integration with respect to angular variables will give zeroes; we shall discuss it below), we find

$$\overset{\bullet}{Q} = (ZT_0)^{-1} \int \tau_p\langle f_p\rangle \left\{ (v\nabla T)^2 \frac{[\varepsilon_p^2 + (pV)^2]}{T_0^2} + [(p\nabla)(vV)]^2 \right.$$

$$\left. - \frac{2(v\nabla T)(pV)(p\nabla)(vV)}{T_0} \right\} \mathrm{d}^3 p \,. \tag{5.157}$$

Here, for further simplification of the obtained dissipative function, we should use the rule of angular variables integration (averaging with respect to angles), which we have already used above. Indeed, we take $\langle v_i v_k \rangle = (\delta_{ik}/3)v^2$, $\langle v_i v_k p_n p_m \rangle = v^2 p^2 a(\delta_{ik}\delta_{nm} + \delta_{in}\delta_{km} + \delta_{im}\delta_{kn})$, where the coefficient "a" is to be determined.

This simple procedure is performed by convolution over indices $i - k$ and $n - m$ of the left and right terms of the equality. In the result $v^2 p^2 = v^2 p^2 a(3 \bullet 3 + 3 + 3) = 15 v^2 p^2$ and hence $a = 1/15$.

Thus, the averaging rule for the four term product leads to

$$\langle v_i v_k p_n p_m \rangle = \left(\frac{v^2 p^2}{15}\right) (\delta_{ik}\delta_{nm} + \delta_{in}\delta_{km} + \delta_{im}\delta_{kn}).$$

Due to the described procedure, Eq. (5.156) allows us to present the dissipative function as

$$\dot{Q} = (ZT_0)^{-1} \times \int \tau_p \langle f_p \rangle \left\{ \left(\frac{v^2 \varepsilon_p^2}{3T_0^2}\right) (\nabla T)^2 + \left(\frac{p^2 v^2}{15}\right) \right.$$

$$\times \left[(\text{div}\boldsymbol{V})^2 + \left(\frac{\partial V_k}{\partial x_i}\right)^2 + \left(\frac{\partial V_k}{\partial x_i}\right)\left(\frac{\partial V_i}{\partial x_k}\right) \right] - \left(\frac{2p^2 v^2}{15T_0}\right)$$

$$\left. \times [(\boldsymbol{V}\nabla T)\text{div}\boldsymbol{V} + \nabla T(\boldsymbol{V}\nabla)\boldsymbol{V} + \boldsymbol{V}(\nabla T\nabla)\boldsymbol{V}] \right\} d^3 p. \tag{5.158}$$

By means of the obtained equation we can now relate the permeability coefficient K, figuring in Darcy's law, to the dissipative characteristics of the stream of liquid being filtered.

Indeed, since (see general Eq. (5.131)) by order of magnitude

$$\int \boldsymbol{F} dV \cong \boldsymbol{F}\boldsymbol{V} = \dot{Q},$$

then, neglecting in (5.158) the non-uniformity of the stream velocity on the scales of the membrane thickness δ, we may take all partial derivatives from the velocity equal to zero, i.e. $\partial V_i / \partial x_k = 0$. Then we have

$$\boldsymbol{F}\boldsymbol{V} = \left(\frac{1}{3}ZT_0\right) \int \tau_p v^2 \langle f_p \rangle \{\varepsilon_p^2 (\nabla T)^2 + 0.2 p^2 [(\nabla T)^2 V^2 + 2(\boldsymbol{V}\nabla T)^2]\} d^3 p.$$

$$\tag{5.159}$$

Force F in the left-hand side of the above equation may be estimated in the following way. Since $\boldsymbol{F} = \boldsymbol{S}(P_1 - P_2) = -\delta |\boldsymbol{S}|\nabla P$, where δ is the membrane thickness and $|\boldsymbol{S}| = S$ is the percolation area, then $\boldsymbol{F}\boldsymbol{V} = -\delta S \nabla P V$. But, on the other hand, according to Darcy's law, the filtration (percolation) velocity is $\boldsymbol{V} = -K\nabla P$, and hence the dissipative function will be equal to

$$\dot{Q} = \delta S K (\nabla P)^2. \tag{5.160}$$

Equating (5.159) and (5.160), we find the following equation:

$$\delta SK(\nabla P)^2 = \left(\frac{1}{3}ZT_0\right)\int \tau_p v^2 \langle f_p\rangle$$
$$\times \{\varepsilon_p^2(\nabla T)^2 + 0.2p^2[(\nabla T)^2 V^2 + 2(\boldsymbol{V}\nabla T)^2]\}\mathrm{d}^3 p. \qquad (5.161)$$

And now it is high time to remember the Joule–Thomson effect, according to which changing the temperature and pressure on both sides of the partition satisfies the relation $\delta T = \alpha_W \delta P$. If we rewrite this equality in terms of gradient operator, we get $\nabla T = \alpha_W \nabla P$. Substituting this relation into the right-hand side of (5.161) and considering also that the velocity has only the component V_z, not equal to zero and perpendicular to the membrane plane, viz. $\boldsymbol{V} = (0, 0, V_z)$ (the plane x–y lies in the membrane plane), and the temperature gradient has components $\nabla T = (0, 0, \partial T/\partial z)$, we find the desired expression for the permeability coefficient as

$$K = \left(\frac{\alpha_W^2}{3\delta ZST_0^3}\right)\int \tau_p v^2 \langle f_p\rangle(\varepsilon_p^2 + 0.6p^2 V_z^2)\mathrm{d}^3 p. \qquad (5.162)$$

So, we have shown that the filtration velocity is connected with the entropy of the liquid (more exactly with its time derivative!!) and is defined by friction in the membrane pores. The entropy, in turn, is the magnitude proportional to the logarithm of the phase volume, that is $S = \ln \Delta\Gamma$, where $\Delta\Gamma$ is the volume of phase space. Hence it follows that the entropy derivative will be $\mathrm{d}S/\mathrm{d}t = (\Delta\Gamma)^{-1}\mathrm{d}(\Delta\Gamma)/\mathrm{d}t$. According to Liouville's theorem, if collisions are absent phase, the space-volume $\Delta\Gamma$ is conserved and $\Delta\Gamma = \Delta\Gamma_{\max}$, and therefore $\mathrm{d}S/\mathrm{d}t = 0$ and hence $S = S_{\max} = \ln(\Delta\Gamma_{\max})$. Taking account of the irreversibility of the filtration process allows us to state that $\mathrm{d}(\Delta\Gamma)/\mathrm{d}t > 0$: the phase-space volume taking account of collisions grows and at $t \to \infty$ tends to occupy the largest volume $\Delta\Gamma_{\max}$! That is, in non-equilibrium conditions the phase volume at the initial moment of time $t = 0$ decreases and then for time $t > 0$ increases, reaching the highest possible value. Due to the presence of the molecular stream entrainment velocity V, the Maxwell distribution function becomes a quasi-equilibrium and one contains in addition to the kinetic energy of the translation motion $\varepsilon_p = p^2/2M$ the addend pV (see above). Thus, the entropy S begins to depend on the stream velocity V. That is why the velocity derived from the dissipative function will give us the desired friction force $F = F_{\tau p}$. Mathematically, all this may be expressed as $F = \partial(T\mathrm{d}S/\mathrm{d}t)/\partial V$, which was used above in order to find the relation between the permeability coefficient and the microscopic properties of the liquid (5.162).

If we return to the formula (5.162) and use the gas-kinetic approximation, according to which the coefficient of kinematic viscosity ν can be estimated by the relation $\nu \approx Z^{-1}\int \tau_p v^2 \langle f_p\rangle \mathrm{d}^3 p$, we can easily estimate the relation between K and ν. Actually, taking $\varepsilon_p \sim T_0$ and $v_T \gg V_z$ (the latter inequality allows to neglect the addend $0.6p^2 V_z^2$ as compared to ε_p^2), we find

$$K = \frac{\alpha_W^2 \nu}{3\delta S T_0}. \tag{5.163}$$

Actually, the obtained formula (5.163) reflects the solution of the problem of finding the relation between the permeability K and the Joule–Thomson coefficient α_W.

But that is not all. The fact is that in reality we wanted to relate α_W to the dissipative properties of the membrane (partition), i.e. to friction. The given equation just copes with this problem. Indeed, expressing from it (from (5.163)) α_W, we find $\alpha_W = (3KV_P T_0/\nu)^{1/2}$, where $V_P = S\delta$ is the volume of the pores. If now we use the dependence which we know from Sect. 5.8 (see (5.124) and the following text) $K = A/\eta$, where A may, for example, be chosen in the form of the relation obtained by Slichter, that is $A_S = m^2 d^2/96(1-m)$, where d in Slichter's model is the diameter of closely packed cylinders, we shall find, if we introduce the liquid density ρ, that the real Joule–Thomson coefficient may be estimated by the following dissipative relation:

$$\alpha_W = \left(\frac{md}{4\nu}\right)\left[\frac{V_P T_0}{2\rho(1-m)}\right]^{1/2}. \tag{5.164}$$

Indeed, in the case of an ideal liquid (or gas) the interaction between the molecules is equal to zero and consequently the relaxation time $\tau \to \infty$, but the kinematic viscosity $\nu = \tau v_T^2$. Therefore $\nu \to \infty$ and hence $\alpha_W = 0$, which should be the case in reality.

6 Behavior of Porous Dielectrics in Acoustic and Electromagnetic Fields

In this chapter we are moving on to analyze very fine physical effects resulting from the investigation of the influence of external acoustic and electromagnetic fields on porous structures.

It should be noted that the range of problems outlined and described in the previous text of the monograph could not claim to be complete and to present the whole picture (collecting together all the main properties of porous media) without paying proper attention to Chap. 6.

To begin, let us consider one concrete example of external action and carry out a small qualitative analysis of it.

Thus, let monochromatic acoustic wave of frequency? fall on a porous dielectric. As we know from a general course of physics, any sound action leads to the deformation of any medium exposed to this action. If the medium is homogeneous (porosity m is equal to zero), then there are macroscopic deformational changes in the dielectric volume, connected merely with travel of points of the continuum medium. Description of its properties may be easily made by means of kinetic equations of evolution of the phonon distribution function. Here two possibilities appear: either the frequency ω is high or it is low. It is most convenient to compare the frequency (see below) with the inverse relaxation time of phonons τ_p. Then, if the frequency is high ($\omega\tau_p \gg 1$), the calculation of attenuation time of the external sound may be performed in terms of the Boltzmann quasi-classical equation. If the frequency is low ($\omega\tau_p \ll 1$), the Boltzmann equation does not "work", and what does begins "to work" is a purely classic description of the acoustic wave attenuation process connected with viscosity and thermal conductance of the medium which is possible at the hydrodynamic limit. Sound attenuation $\gamma(\omega)$ for low frequencies from the range $\omega\tau_p \ll 1$, will be proportional to the squared frequency ω, as it should be according to the classical result.

Thus, everything seems to be clear with the homogeneous structure. Suppose now, that we have changed the composition of the homogeneous dielectric by introducing into it only one single pore of radius R. What will happen? If the frequency of the acoustic field ω is not high and its wavelength exceeds the pore radius R, then in terms of external sound such a medium may be treated as a homogeneous continuum and the description of its properties may be easily made in accordance with the above technique. But if the wavelength is small compared to R, such a situation appears to be the most interesting for analyzing. In this case it is quite clear that the solution of the problem should be connected with the account of the absorption effect of the external sound wave on the boundary of the pore volume V_0. This is one contribution. The other contribution is connected with the acoustic signal relaxation in the rest region $V - V_0$, if V implies the dielectric volume. The result of scattering, according to the well-known theorem from probability theory on independent and incompatible events, should be the sum of these two contributions. Here one part connected with the scattering on the pore boundary will, in addition, include the radius R! The situation is very curious.

Now it is time to discuss directly our specific fibril-porous structures, with which we deal throughout the Monograph, and the present chapter will be no exception!

In investigating energy absorption processes (or intensity) for an external alternating field the relation between the characteristic relaxation time τ and the field frequency ω is of great importance. The character of the absorption, naturally, will greatly vary, depending on the structure and composition of the substance. For a comprehensive study we shall choose again a porous dielectric and from this example we shall observe why and how the absorption pattern and the main dissipative characteristics of such structures (dielectric permeability, scattering cross-section, attenuation coefficients, etc.) change.

Adhering to the concept of the fibril structure we shall consider two phases – gaseous and solid – and for this reason introduce the time τ_1 corresponding to three-particle and four-particle processes of phonon-phonon scattering in fibrils, and the time τ_2 characterizing binary molecule collisions in the gaseous phase. We shall assume that either the time is the smallest in its phase (which is confirmed by direct estimations, as well). Also, let time τ_1' correspond to some abstract (so far!) mechanism of interaction between the phonons and the thermostat (a thermal reservoir, often called a "bath"), and the time τ_2' be the time of approaching equilibrium in a molecular subsystem in the gas. Their hierarchy is such: $\tau_1, \tau_2 \ll \tau_1', \tau_2'$. Let us now analyze how the qualitative pattern of absorption will change when the frequency of the alternating action ω falls into different intervals on the τ-scale.

1. Let the frequency satisfy the inequality

$$\tau_1'^{-1}, \quad \tau_2'^{-1} \ll \omega \ll \tau_1^{-1}, \quad \tau_2^{-1}. \tag{6.1}$$

If the given inequality holds, we may speak about weak withdrawal of systems (molecules in the gaseous phase and phonons in the fibrils) from the state of thermodynamic equilibrium which is due to the external alternating field. Indeed, during the time τ_1 and τ_2 in both phases quasi-equilibrium states are reached with temperatures equal to T_1 and T_2 respectively, and then due to the connection with the thermostat (whose function may be performed, for example, by phonons), these temperatures will relax to the single equilibrium temperature T. The respective relaxation time intervals are τ_1' and τ_2'. It is the connection with the thermostat which causes the whole absorption effect. In dielectrics this absorption is defined as the imaginary term of the dielectric permeability tensor, which we denote as $\varepsilon''(\omega)$. If we deal with a two-phase medium, as in this case, characterized as well by porosity the m, then $\varepsilon''(\omega)$ is determined from the sum of the magnitudes $\varepsilon_1''(\omega)$ and $\varepsilon_2''(\omega)$ and the equation

$$\varepsilon''(\omega) = (1 - m)\varepsilon_1''(\omega) + m\varepsilon_2''(\omega)\,, \tag{6.2}$$

where $\varepsilon_1''(\omega)$ and $\varepsilon_2''(\omega)$ are the permeability of the solid and gaseous phases, respectively.

For the Debye frequency dependence, for which

$$\varepsilon_{1,2}''(\omega) = \frac{\varepsilon_0 \omega \tau_{1,2}'}{(1 + \omega^2 \tau_{1,2}'^2)}\,,$$

we have

$$\varepsilon''(\omega) = \frac{(1 - m)\varepsilon_{01}\omega\tau_1'}{(1 + \omega^2\tau_1'^2)} + \frac{m\varepsilon_{02}\omega\tau_2'}{(1 + \omega^2\tau_2'^2)}\,, \tag{6.3}$$

where ε_{01} and ε_{02} are the static-phase dielectric permeabilities.

Due to condition (6.1) we have in this case

$$\varepsilon''(\omega) = \frac{(1 - m)\varepsilon_{01}}{\omega\tau_1'} + \frac{m\varepsilon_{02}}{\omega\tau_2'} \tag{6.4}$$

or

$$\varepsilon''(\omega) = \frac{\varepsilon_{0ef}}{\omega\tau_{0ef}}\,,$$

where

$$\frac{\varepsilon_{0ef}}{\tau_{0ef}} = (1 - m)\frac{\varepsilon_{01}}{\tau_1'} + \frac{m\varepsilon_{02}}{\tau_2'}\,. \tag{6.5}$$

2. Let the frequency now lie in the range

$$\omega \ll \tau_1'^{-1}\,, \quad \tau_2'^{-1} \ll \tau_1'^{-1}\,, \quad \tau_2'^{-1}\,. \tag{6.6}$$

It is clear that in this case for a large period of the field oscillations $T_0 = 2\pi/\omega$ the particle and quasi-particle systems will have time to approach the true equilibrium state. The absorption in the framework of the general formula (6.3) and including (6.6) is described by the dependence:

$$\varepsilon''(\omega) = (1 - m)\varepsilon_{01}\omega\tau_1' + m\varepsilon_{02}\omega\tau_2'\,. \tag{6.7}$$

3. Suppose now that the frequency ω satisfies the inequality:

$$\tau_1'^{-1} \ll \omega \ll \tau_2'^{-1} \ll \tau_1'^{-1}\,, \quad \tau_2'^{-1}\,. \tag{6.8}$$

Then it is obvious that during a field period one of phases (molecules) will manage to come to equilibrium with the thermostat due to the condition $\omega\tau_2'' \ll 1$, and the other phase – the phonon medium – will not, because

$\omega\tau_2'' \gg 1$. This, in turn, means that the imaginary part of the permeability is

$$\varepsilon''(\omega) = \frac{(1-m)\varepsilon_{01}}{\omega\tau_1'} + m\varepsilon_{02}\omega\tau_2'. \tag{6.9}$$

In this relation the function $\varepsilon''(\omega)$ passes through a minimum defined by the frequency

$$\omega_0 = \left[\frac{(1-m)\varepsilon_{01}}{m\varepsilon_{02}\tau_1'\tau_2'}\right]^{1/2},$$

$$\varepsilon''_{\min}(\omega) = \left[\frac{m(1-m)\varepsilon_{01}\varepsilon_{02}\tau_2'}{\tau_1'}\right]^{1/2}. \tag{6.9a}$$

It should be noted that such a specific absorption minimum arises only in polydispersed systems. Indeed, if we set $m = 0$, we get $\varepsilon''_{\min}(\omega) = 0$ too.

Exchanging the positions of τ_1' and τ_2' in the inequality (6.9) we find

$$\varepsilon''(\omega) = (1-m)\varepsilon_{01}\omega\tau_1' + \frac{m\varepsilon_{02}}{\omega\tau_2'} \tag{6.10}$$

and, respectively,

$$\varepsilon''_{\min}(\omega) = \left[\frac{m(1-m)\varepsilon_{01}\varepsilon_{02}\tau_1'}{\tau_2'}\right]^{1/2}. \tag{6.10a}$$

Now some words about the last and the most complicated case.

4. Frequency

$$\omega \gg \tau_1'^{-1}\tau_2'^{-1} \gg \tau_1^{-1}, \quad \tau_2^{-1}. \tag{6.11}$$

At frequencies in the indicated range both phases are short of time to reach the quasi-equilibrium state, and in this case for small times ω^{-1} we have no right to introduce such a notion as temperature (though in particular specific cases it can be done; see [54])! It means that the linear theory for studying the system's response to external action (see Chap. 5, Kubo's formula) does not "work", and other ways to solve the problems should be sought for. Such an approach, in particular, was described in the paper [54] already repeatedly mentioned. Without dwelling upon it in detail we shall only note that in this case ε'' begins to depend not only on the frequency ω, but on the amplitude of the applied field.

Now we shall pass over to concrete problems.

6.1 Energy Absorption Coefficient of the External Low-Frequency Electromagnetic Field of Porous Structures

For dealing with the statistical ensemble of a microscopic or macroscopic system, no problem can be solved without such a powerful investigation tool as the kinetic equation for the corresponding distribution function. In most problems, however, the corresponding integral-differential equation is nonlinear and its exact solution involves sufficient difficulties of a purely calculation character which can be overcome only by approximate methods or computer calculation. It is obvious that only those equations may be exceptions from the rules whose "collision integrals" have a linear dependence on the distribution function. In many cases such equations may be confined to the differential equation of Fokker–Planck or Einstein–Smoluchowskii (see, for example, monograph [105]), which in principle is solved purely analytically in the framework of a concrete problem with the corresponding initial and boundary conditions. If we depart from the linear (in distribution function) "collision integral", its solution in some particular cases also may be found analytically by means of the method of expansion of the distribution function into a Fourier integral (see Sect. 4.2, paper [38], and monograph [112]). However, it is interesting to note that in the overwhelming number of cases when we deal with the calculation of, say, the thermal conductivity coefficients, or the diffusion or some other parameters of the internal non-equilibrium state of the structure under investigation, it will be sufficient to restrict ourselves to the so-called τ-approximation. The latter is good because it allows us to linearize the collision operator and thus to simplify the corresponding transformations. Besides, as can be seen in practice, the τ-approximation has wide applications, not only to explain the available experimental data on non-equilibrium in systems but also even to predict a number of qualitatively new physical effects.

The problem we are going to solve belongs to the category of problems referred to as the τ-approximation. It concerns the calculation of the temperature dependence of the dielectric susceptibility α of porous dielectrics placed into an alternating electric field $\boldsymbol{E'}(t) = \boldsymbol{E'_0} e^{-i\omega t}$ for small frequencies ω. The calculation of the imaginary part $\alpha(\omega, T)$ is somewhat complicated by the fact that it is not clear which particular relaxation mechanism will play the main role. However, let us try to outline the methods of solution, relying upon purely qualitative reasoning. As we deal with the ultra-low range of frequencies (we shall consider only frequencies in the range of about 10^2–10^3 Hz) for which it is always possible to find such a relaxation time τ that the inequalities $\omega\tau \ll 1$ and $\omega\tau \gg 1$ will apply, the conclusion suggests itself that the main contribution to $\alpha''(\omega)$ will be made by the macroscopic relaxation time connected with the relaxation of the polarization (or dipole moment). Such a not quite rigorous explanation of the macroscopic nature of the origin of the time τ will be reasonable only in the case of its rigorous analytical calculation followed by numerical estimation. Only the estimation will be able to show the correctness of the chosen method to solve the problem. But before we attend to the problem we shall say some words about the model representation of the dielectric internal structure. According to the model of f-points (see Chap. 2) we shall consider the composition of the substance as an interlacing of fibrils and free volumes (pores), as shown in Fig. 6.1. The unhatched areas correspond to the gaseous phase and the hatched ones to

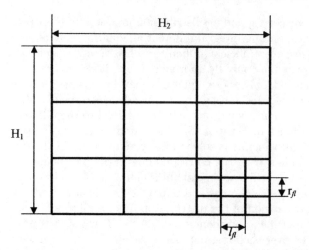

Fig. 6.1. "Chess" scheme of regions with different conductivity. The total area is $S = H_1 \times H_2$, and the dimensions of each "square" are l_{fl} and r_{fl}.

fibrils. Note, by the way, that it is for such a structure that (6.2) will be valid. Remember also that the relation between the dielectric permeability ε and the susceptibility α is very simple: $\varepsilon = 1 + 4\pi\alpha$.

Besides, we have

$$\varepsilon''(\omega) = 4\pi\alpha''(\omega). \tag{6.12}$$

The determination of α is entirely connected with the frequency ω. The thing is that when we deal with high frequencies, for adequate clarification of the dependence $\alpha(\omega, T)$ it is necessary to distinguish strictly between absorption in the fibrils and in the gaseous phase because their relaxation occurs in different ways. In the case of the low frequencies which are under discussion, such a distinction is smoothed over and we may introduce a single relaxation time of average (over the dielectric volume) polarization. Indeed, in this case the dipole moment localized in any point of the volume can follow the time variation of the electric intensity $\boldsymbol{E}'(t)$.

Thus, to solve the problem we shall use a phenomenological approach, and for this purpose we shall introduce a distribution function of the dipole moments $\varphi(\theta, t)$, where the angle θ is defined from the z axis along which we shall direct the electric field $\boldsymbol{E}_0$.

We have

$$\frac{\partial\varphi(\theta, t)}{\partial t} = 2\pi_{k,\theta'}\sum|A_k|^2\{[\varphi(\theta', t)(1 + \langle n_k\rangle) - \varphi(\theta, t)\langle n_k\rangle]$$
$$\times \delta[dE_0(\cos\theta - \cos\theta') - \hbar\omega_{pk}]$$
$$+ [\varphi(\theta', t)\langle n_k\rangle - \varphi(\theta, t)(1 + \langle n_k\rangle)]$$
$$\times \delta[dE_0(\cos\theta - \cos\theta') + \hbar\omega_{pk}]\}, \tag{6.13}$$

where the amplitude

$$A_k = i\xi \left[\frac{\hbar k}{2\rho c_s(m)V} \right]^{1/2} , \qquad (6.14)$$

ξ is the phenomenological constant of the phonon interaction with the dipole moment, $V = S\delta$ is the dielectric volume, S is the area of its surface and ρ is the structure density. The sound velocity as a function of the porosity m was introduced in Chap. 3 (see formulae (3.9)–(3.12)). Having slightly modified the formula (3.11) we shall present it as follows:

$$[\rho c_s^2(m)]^{-1/2} = (1 - m)(\rho_1 c_{1s}^2)^{-1/2} + m(\rho_2 c_{2s}^2)^{-1/2} , \qquad (6.15)$$

where c_{1s} and c_{2s} are, respectively, the sound velocities in the solid phase (in the fibrils) and in the gaseous phase (in the pores).

Further, the equilibrium function of the phonon distribution is

$$\langle n_k \rangle = \left(\exp \left\{ \frac{\hbar \omega_{pk}}{T} \right\} - 1 \right)^{-1}$$

and their dispersion (sound frequency) is $\omega_{pk} = c_s(m)k$.

From (6.13) it follows that in the equilibrium state the distribution function $\langle \varphi \rangle$ is

$$\langle \varphi \rangle = R \exp \left\{ -\frac{dE_0 \cos \theta}{T} \right\} , \qquad (6.16)$$

where the normalization constant is

$$R = \frac{dE_0}{Tsh} \left(\frac{dE_0}{T} \right)^{-1} . \qquad (6.17)$$

In the region of high temperatures $T \gg \hbar\omega_{pk}$, (6.13) will be simpler. With the view to reduce writing we shall omit the explicit dependence of the classic distribution function φ on the angle and get

$$\frac{\partial \varphi(\theta, t)}{\partial t} = \nu \sum_k \sum_{\theta'} [\varphi(\theta', t) - \varphi(\theta, t)] \{ \delta[dE_0(\cos\theta - \cos\theta') - \hbar\omega_{pk}]$$
$$+ \delta[dE_0(\cos\theta - \cos\theta') + \hbar\omega_{pk}] \} , \qquad (6.18)$$

where

$$\nu = \frac{\pi \xi^2 T}{\rho c_s^2(m)V} . \qquad (6.19)$$

The equation (6.18) may be reduced to a differential one if $\theta' = \theta + \delta\theta$, where $\delta\theta$ is small compared to θ. Indeed, in this case we may expand the function $\varphi(\theta') = \varphi(\theta + \delta\theta)$ into a power series in $\delta\theta$. As a result we find

$$\frac{\partial \varphi}{\partial t} = \alpha \frac{\partial \varphi}{\partial(\cos\theta)} + \beta \frac{\partial^2 \varphi}{\partial(\cos\theta)} (\cos\theta) . \qquad (6.20)$$

The stationary solution of this equation gives the equilibrium distribution function (6.16) on condition that $\beta = \alpha T/\mathrm{d}E_0$.

The solution of (6.20) allows us to find the non-stationary behaviour of the distribution function. The sphere of applicability of (6.13) and (6.20) will be discussed later.

In order to calculate the relaxation time of the local dipole moment whose change of direction is due to phonon absorption (radiation) processes we shall return to (6.18) and find the functional derivative of the right-hand side with respect to φ. Then

$$\tau_\theta^{-1} = -\frac{\delta(\partial\varphi/\partial t)}{\delta\varphi|_{\varphi=\langle\varphi\rangle}}.$$

Substituting here the right-hand side of (6.18) and passing from summation over k to integration, which is done by the replacement

$$\sum_\kappa(\ldots) = V^{-1} \int \mathrm{d}^3 k(\ldots),$$

after which with (6.19) we finally find

$$\tau_\theta^{-1} = \frac{(2\xi^2/\pi\hbar)[\mathrm{d}E_0/\hbar c_s(m)]^2 T}{[\rho c_s^2(m)(\cos^2\theta + 1/3)]}. \tag{6.21}$$

Now, in order to calculate the dielectric susceptibility of a sample it is necessary to look for a solution of the form:

$$\varphi(\theta, t) = \langle\varphi\rangle + \delta\varphi(\theta, t). \tag{6.22}$$

Attention should be paid once more to the fact that we are looking for the solution of (6.18) in the τ-approximation, thereby demonstrating the capabilities of such an approach as applied to various problems.

Since the energy of the dipole interaction of a polarizing atom with the alternating field E' can be presented as

$$\varepsilon = d[E_0 + \alpha_0 E'(t)],$$

where α_0 is the static polarizability of an atom (remember, that we are dealing with very low frequencies), then using (6.22) the Eq. (6.18) gives

$$\left(\frac{\partial\langle\varphi\rangle}{\partial\varepsilon}\right)\alpha_0 dE'(t) + \frac{\partial\delta\varphi}{\partial t} = -\frac{\delta\varphi}{\tau_\theta}. \tag{6.23}$$

We are trying to solve the obtained equation in the form

$$\delta\varphi(\theta, t) = \delta\varphi(\theta, 0)\mathrm{e}^{-\mathrm{i}\omega t}.$$

Then after arithmetic transformations we immediately find

$$\delta\varphi(\theta, t) = \left[\frac{\mathrm{i}\omega\tau_\theta\alpha_0 dE'(t)}{(1 + \mathrm{i}\omega\tau_\theta)}\right]\frac{\partial\langle\varphi\rangle}{\partial\varepsilon_0}, \tag{6.24}$$

where φ is given by (6.16) and $\varepsilon_0 = dE_0$.

Then, as the dipole moment averaged over the Gibbs macroensemble is

$$\boldsymbol{d}(t) = \boldsymbol{d} \int\limits_0^\pi \cos\theta \varphi(\theta, t) \sin\theta \mathrm{d}\theta \,,$$

then after substitution here of (6.22) and with (6.24) we find

$$\boldsymbol{d}(t) = \langle\boldsymbol{d}\rangle + \alpha(\omega)\boldsymbol{E}'(t)\,, \qquad (6.25)$$

where $\boldsymbol{d} = dL(dE_0/T)$ and L is Llangowen's function

$$L(x) = \mathrm{cth}\,x - \frac{1}{x}\,. \qquad (6.26)$$

The dielectric susceptibility coefficient sought for is determined by the equation

$$\alpha(\omega) = \left(\frac{\mathrm{i}\omega\alpha_0 d^2}{V}\right) \int\limits_0^\pi \left[\frac{\tau_\theta \cos\theta}{(1 + \mathrm{i}\omega\tau_\theta)}\right] \left(\frac{\partial\langle\varphi\rangle}{\partial\varepsilon_0}\right) \sin\theta \mathrm{d}\theta\,. \qquad (6.27)$$

Using relations (6.16), (6.17), and (6.21) we shall rewrite the sought-for dependence for $\alpha(\omega, T)$ in the form:

$$\alpha(\omega, T) = \alpha'(\omega, T) + \mathrm{i}\alpha''(\omega, T)\,,$$

where the real and imaginary parts are

$$\begin{cases} \alpha'(\omega, T) = \dfrac{\mathrm{d}^3 E_0 \omega^2 \tau_0^2 J_1(\omega)}{T^2 V \, sh(dE_0/T)}\,, & (6.28) \\[4mm] \alpha''(\omega, T) = \dfrac{\mathrm{d}^3 E_0 \omega \tau_0 J_2(\omega)}{T^2 V \, sh(dE_0/T)}\,, & (6.29) \end{cases}$$

The functions J_1 and J_2 are given by

$$\begin{cases} J_1(\omega) = \displaystyle\int\limits_{-1}^1 \dfrac{x e^{-ux} \mathrm{d}x}{(x^2 + 1/3)^2 + (\omega\tau_0)^2}\,, & (6.30) \\[6mm] J_2(\omega) = \displaystyle\int\limits_{-1}^1 \dfrac{x(x^2 + 1/3)e^{-ux} \mathrm{d}x}{(x^2 + 1/3)^2 + (\omega\tau_0)^2}\,, & (6.31) \end{cases}$$

where parameter $u = dE_0/T$ and the inverse time is

$$\tau_0^{-1} = \frac{2\xi^2 (dE_0)^2 T}{\pi\hbar(\hbar c_s(m))^3 \rho c_s^2(m)}\,. \qquad (6.32)$$

Unfortunately, it is impossible to calculate the integrals J_1 and J_2 and that is why we shall present only their asymptotic expressions:

$$J_1(u) = \begin{cases} \dfrac{u}{(\omega\tau_0)^2} & \text{at} \quad u \ll 1 \\[2ex] u^{-2}\left[(\omega\tau_0)^2 + \dfrac{1}{9}\right]^{-1} & \text{at} \quad u \gg 1, \end{cases} \tag{6.33}$$

$$J_2(u) = \begin{cases} \dfrac{28u}{45}(\omega\tau_0)^2 & \text{at} \quad u \ll 1 \\[2ex] (3u^2)^{-1}\left[(\omega\tau_0)^2 + \dfrac{1}{9}\right] & \text{at} \quad u \gg 1. \end{cases} \tag{6.34}$$

In order to obtain a final expression for α we should take into account one more relaxation mechanism of the dipole moments. We mean the mechanism of the fluctuation change of the fibril dimensions on periodic exposure to an alternating field $E'(t)$. In Sect. 6.3 this relaxation time will be shown to be

$$\tau_v^{-1} = \frac{v_{fl}\mu^2(T_b - T)}{\kappa P_0(v_{fl} - \langle v \rangle)}, \tag{6.35}$$

where V_F is the fibril volume, $\langle v \rangle$ is the average statistical fibril volume, T_b is the burning temperature, $\mu = 1/s_0\tau_M$ is the coefficient of convection heat transfer from a fibril into the free volume, s_0 is the average area of a fibril surface, τ_M is the time between molecular collisions with the fibril surface, κ is their thermal conductivity coefficient and P_0 is "the pressure" in the fibrils.

Taking account of the time (6.35) means a simple substitution of Eq. (6.32) into $\tau_{ef}^{-1} = m\tau_0^{-1} - (1-m)\tau_v^{-1}$. Let us estimate the time τ_0. Assuming, for example, that $\xi = 10^{-17}$ erg, $d = 10^{-16}$(CGS), $E_0 = 10^2$ V/cm, $T = 300$ K, $c_s(m) = 10^5$ cm/s and $\rho = 1.3$ g/cm^3 we find that $\tau_0 = 10$ s. The time τ_v is $\tau_v = 10^{-4}$ s, and consequently, according to (6.9) and (6.10) the dielectric loss tangent may be estimated by the formula:

$$tg\delta = \frac{\varepsilon''}{\varepsilon'} = \omega\tau_v + \frac{1}{\omega\tau_0}. \tag{6.36}$$

The plot of the temperature dependence of this function at fixed frequency ω will have the shape shown in Fig. 6.2. It should be noted (see this figure) that with the increase of dielectric density $tg\delta$ grows. In fact the problem stated in the beginning of this section may be considered to be solved by the formula (6.36), and we have only to find applicability boundaries of the Eq. (6.13). The first condition on the time interval of the distribution function is obvious:

$$\tau_\theta\delta\theta \ll \delta t \ll \tau_\theta, \tag{6.37}$$

where $\delta\theta$ is some small change of the rotation angle of the dipole moment connected with the phonon scattering. As follows from the law of energy conservation (delta-function argument in the collision integral of (6.13)), the following condition must be satisfied:

$$dE_0\delta\theta|\sin\theta| \geq \delta\varepsilon \Rightarrow \frac{\hbar}{\tau_\theta}. \tag{6.38}$$

$tg\delta \times 10^3$

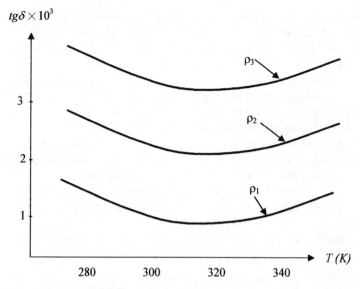

Fig. 6.2. Dependence of the tangent of the loss angle on the temperature and density of the substance

Consequently, the relation (6.37) defines the time interval in which Eq. (6.13) is valid, and the relation

$$\frac{\hbar}{\tau_\theta}|\sin\theta|dE_0 \ll \delta\theta \ll \theta \qquad (6.39)$$

defines the range over which the "coordinate" θ changes.

In case of small rotation angles, as follows from (6.39) at $\sin\theta \cong \theta$, we have

$$\theta \gg \left(\frac{\hbar}{\tau_{\theta=0}dE_0}\right)^{1/2}. \qquad (6.40)$$

Now we come to the last point that with certain assumptions the phenomenological Eq. (6.13) may be reduced to a diffusion equation of the Einstein–Smoluchowskii type (see (6.20)), which describes the change of the dipole moment direction connected with diffusion. The condition necessary for it is the smallness of the angle $\delta\theta$ change connected in general with some fast-oscillating random force $\boldsymbol{F}(t)$. Indeed, the diffusion coefficient is related to it by the equation:

$$D = \langle\boldsymbol{F}(t)\boldsymbol{F}(t')\rangle = \langle\boldsymbol{F}_0^2\rangle\delta(t-t'),$$

where F_0 is the force amplitude. Approximately the same interpretation is characteristic for the Eq. (6.13). Really, in calculating the relaxation time τ_θ it was considered that the correlation time (the time between phonon collisions) τ_{ph} is much less than τ_θ and this, in turn, allowed us to introduce the equilibrium function of the phonon distribution $\langle n_{k,}\rangle$. Besides, as we assured ourselves above, at high temperatures ($T \gg \hbar\omega_{pk}$) the inverse relaxation time τ_θ^{-1} is proportional to T. But just the same dependence is characteristic of the diffusion coefficient, for according to the general theory of fluctuations [11] $\langle F_0^2 \rangle \Rightarrow T$. The described interrelation of both approaches shows the reasonableness of the introduction of the phenomenological Eq. (6.13), which allowed us to calculate the average relaxation time of the dipole moment in a porous medium as a function of the angle θ. This, in particular, shows the advantage of the given calculation method, because in the framework of the diffusion approximation the analogous calculations would be more complicated.

In addition, attention should be paid to the fact that the above-suggested theoretical treatment of the relaxation process of the polarization vector (dipole moment) is somewhat different from that given in monographs [113–115], which is not surprising because every author applies his or her "favourite" method of calculation.

The main results obtained may be summarized in the following three propositions.

1. The suggested theory for calculating the dielectric susceptibility of porous substances from the frequency of the alternating field is based on the application of the quasi-classical relaxation equation introduced phenomenologically.
2. By means of this eq. it appears possible to estimate the relaxation time of dipole moments as a function of the rotation angle θ.
3. Two macroscopic relaxation times τ_θ and τ_v have been found which define the energy losses in a low-frequency alternating electric field in porous fibril structures, and this, in turn, allowed us to introduce a purely technical characteristic-tangent delta.

As to τ_v, it will be considered in more detail in Sect. 6.3.

6.2 Sound Absorption by Porous Dielectrics

As we know, the most available and low cost class of porous structures includes various sorts of cellulose, its derivatives, and paper. For a number of reasons the question was not raised about the calculation of the sound attenuation coefficient, which, as we think, is due to the absence of proper technical equipment for experimentation; moreover, it may be caused by the fact that such materials are produced by a well-developed technological cycle, and therefore there is no reason for a more thorough investigation of the various physical regularities. However, to be objective, we should note that beginning from the 1930s a rather fruitful study has been carried out of the various characteristics relating to electric strength, heat and heat conduction in different types of cellulose, paper (including capacitor paper), ceramics and glass.

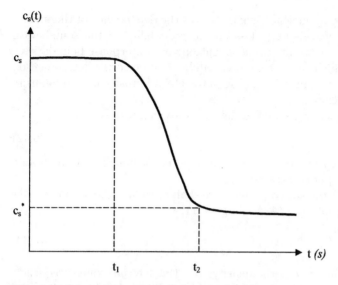

Fig. 6.3. Schematic pattern of the sound velocity versus time

In this connection the theoretical results submitted in the present monograph may first of all serve as an instrument and maybe a prompt for conducting a number of experiments not realized yet but very important for purely practical purposes.

Measurement and analysis of the sound velocity dependence in cellulose (see Table 3.1) which we presented in Sect. 3.1 (see also Sect. 2.5) showed that $c_s(m)$ for small porosity of a substance ($m \ll 1$) abruptly decreases and becomes even lower than the sound wave velocity in a one-phase structure.

Let us now imagine (see Fig. 6.3) that we the measure the time dependence of the intensity of the sound wave energy in a porous substance. Then in the interval of microscopic time $t_1 < t < t_2$ the sound energy density must be described by the equation (quite analogous to the equation of Bouguer–Lambert):

$$E_s(t) = \varepsilon_s + (E_s - \varepsilon_s)e^{-\gamma t}, \qquad (6.41)$$

where γ is the energy loss in unit time (on the definition of energies E_s and ε_s see below).

In principle we may assume that the low concentration of the free volumes is the kind of medium which "attracts" a sound wave and is the only "white spot" on the background of the whole structure. If we adhere to such a concept we shall see that this small concentration of free volumes will lead to the energy loss γ sought for. On the other hand, since $E_s = \rho_1 c_{1s}^2/2$ and $\varepsilon_s = \rho_1 c_s^2(m)/2$, due to the constancy of the densities ρ_1 and ρ_2 it will mean that only $c_s(t)$ changes. On a macroscopic timescale, as it should be (see [11]), the sound velocity in a two-phase medium changes abruptly.

The attenuation coefficient γ characterizes the dependence of the inverse time of the sound wave energy loss or the probability of the sound wave dissipation on the density fluctuations, entropy and interphase boundaries.

The problem which we wish to solve will be reduced to the determination of the coefficient γ as a function of the parameters: porosity m, temperature T and sound frequency ω.

The sound frequency ω satisfies the "hydrodynamic" inequality:

$$\omega \tau_k \ll 1, \tag{6.42}$$

where τ_k is the relaxation time (free path time) of phonons in the solid-body matrix for phonon–phonon scattering.

In order to solve the stated problem we shall introduce the formal expression for the coefficient $\gamma(\omega)$. According to [11]

$$\gamma(\omega) = \frac{\langle \partial E'/\partial t \rangle}{2E_0}, \tag{6.43}$$

where $E_0 = 0.5\rho V \omega^2 u_0^2$, ρ is the density, ω is the external sound frequency, u_0 is the sound wave amplitude, E' is the energy loss due to absorption, and the bar-above means time averaging. The time derivative $\mathrm{d}E'/\mathrm{d}t = T\mathrm{d}S/\mathrm{d}t$, where for the entropy we have the relation:

$$S = (1-m)S_1 + mS_2,$$

where

$$\begin{cases} S_1 = N_1^{-1} \sum_i [(1+n_{1k})\ln(1+n_{1k}) - n_{1k}\ln n_{1k}], \\ S_2 = -N_2^{-1} \sum_i v_{2i} \int n_{2p} \ln n_{2p} \dfrac{\mathrm{d}^3 p}{(2\pi\hbar)^3}, \end{cases} \tag{6.44}$$

where N_1 is the entire number of atoms in the fibril structure, n_1 is the phonon distribution function, n_2 is the molecular distribution function in the gaseous phase, N_2 is the entire number of molecules in the gas, v_2 is the volume of a single pore and the sum in S_2 is taken over all the volumes of gaseous phase.

In order to calculate the absorption let us take, as usual,

$$n_{1,2k} = \langle n_{1,2k} \rangle + \delta n_{1,2}, \tag{6.45}$$

where

$$\begin{cases} \langle n_{1k} \rangle = \left(\exp\left\{ \dfrac{\hbar \omega_k}{T} \right\} - 1 \right)^{-1}, \\ \langle n_{2k} \rangle = Z\mathrm{e}^{-\varepsilon(p)/T}. \end{cases} \tag{6.46}$$

The constant Z is a normalization factor of the equilibrium distribution function for molecules, $\varepsilon(p) = p^2/2M$ and $\omega_k = c_s k$.

Differentiating (6.44), taking account of (6.45) and (6.46), we find for the energy losses:

$$\frac{\mathrm{d}E'}{\mathrm{d}t} = T\frac{\mathrm{d}S}{\mathrm{d}t} = -T(1-m)N_1^{-1}\sum_k \frac{(\mathrm{d}\delta n_{1k}/\mathrm{d}t)\delta n_{1k}}{\langle n_{1k}\rangle(1+\langle n_{1k}\rangle)}$$

$$- mTN_2^{-1}\sum_i v_{2i}\int\left(\frac{\mathrm{d}\delta n_{2p}}{\mathrm{d}t}\right)\left(\frac{\delta n_{2p}}{\langle n_{2p}\rangle}\right)\frac{\mathrm{d}^3p}{(2\pi\hbar)^3}. \qquad (6.47)$$

It should be noted that in the derivation of (6.47) we have omitted terms linear in δn_1 and δn_2. It is due to the fact that in further averaging of the relation for entropy they become zeros.

Then, because changes of the distribution functions satisfy the equation

$$\frac{\mathrm{d}\delta n_{1,2k}}{\mathrm{d}t} = -\delta n_{1,2k/1,2k'} \qquad (6.48)$$

for the average (over the sound wave period) value of energy losses per unit time from the relation (6.47) we find

$$\frac{\mathrm{d}E'}{\mathrm{d}t} = T\frac{\mathrm{d}S}{\mathrm{d}t} = \frac{T(1-m)N_1^{-1}\sum_k\langle\delta n_{1k}\rangle^2}{\tau_{1k}'\langle n_{1k}\rangle(1+\langle n_{1k}\rangle)}$$

$$+ mTN_2^{-1}\sum_i v_{2i}\int\left(\frac{\langle\delta n_{2p}\rangle^2}{\tau_{2p}'\langle n_{2p}\rangle}\right)\left[\frac{\mathrm{d}^3p}{(2\pi\hbar)^3}\right]. \qquad (6.49)$$

In the second sum we take $p = \hbar k$, though it should be noted that it is not quite correct, for we consider molecules as solid balls rather than a wave. It is done for the reason of dimensionality.

Let us begin with the calculation of the correction to the equilibrium function of the phonon distribution due to their quasi-elastic interaction with the density fluctuations.

When a plane sound wave falls on a substance, the phonon spectrum in the solid structure deforms and becomes renormalized, and so the phonon dispersion in the process of changing must be substituted for

$$\omega_k^* = \omega_k(1 + g_{ik}u_{ik}), \qquad (6.50)$$

where g_{ik} is a tensor coefficient of the solid matrix depending on the temperature and the symmetric deformation tensor:

$$u_{ik} = 0.5\left(\frac{\partial u_i}{\partial x_k} + \frac{\partial u_k}{\partial x_i}\right).$$

The final results of the calculations will not be greatly distorted (at least in order of magnitude they will comply with reality), if we assume that the dielectric structure is isotropic and, therefore, the dispersion is

$$\omega_k^* = \omega_k(1 + g\,\mathrm{div}\boldsymbol{u}). \qquad (6.51)$$

Let

$$\boldsymbol{u}(\boldsymbol{x},t) = \boldsymbol{u}_0\mathrm{e}^{\mathrm{i}(\boldsymbol{qx}-\omega t)}. \qquad (6.52)$$

The external sound frequency is related to q by the "common relation" $\omega = c_s q$.

From the above, (6.48) in the expanded form must be written as follows:

$$\frac{\partial \delta n_{1k}}{\partial t} + \left(\frac{\partial \langle n_{1k} \rangle}{\partial \omega_k}\right)|_{g=0} \left(\frac{\partial \omega_k^*}{\partial t}\right) = -\frac{\delta n_{1k}}{\tau'_{1k}}. \tag{6.53}$$

Substituting here (6.51) together with (6.52) we find

$$\frac{\partial \delta n_{1k}}{\partial t} + g c_s(m)(\boldsymbol{nq})(\boldsymbol{uq}) \frac{\omega_k \partial \langle n_{1k} \rangle}{\partial \omega_k} = -\frac{\delta n_{1k}}{\tau'_{1k}}. \tag{6.54}$$

We solve the obtained equation in the form:

$$\delta n_{1k}(t) = \delta n_{1k}(0) e^{i\boldsymbol{qx} - i\omega t}. \tag{6.55}$$

Then from (6.54) we immediately get

$$\delta n_{1k}(0) = -g c_s T(\boldsymbol{nq})(\boldsymbol{uq}) \left[\frac{\partial \langle n_{1k} \rangle}{\partial T}\right] (i\omega - \tau'_{1k})^{-1}. \tag{6.56}$$

In deriving this relation the following identity was taken into account:

$$\frac{\omega_k \partial \langle n_{1k} \rangle}{\partial \omega_k} = -\frac{T \partial \langle n_{1k} \rangle}{\partial T}.$$

From the solution (6.56) the sought-for losses described by the first term in (6.49) may be determined in the form:

$$\text{Re} \left\langle \frac{\partial E'}{\partial t} \right\rangle = g^2 T^3 u_0^2 \omega^4 (1 - m) \sum_k \frac{(\partial \langle n_{1k} \rangle / \partial T)^2 (\tau'^{-2}_{1k} - \omega^2)}{\langle n_{1k} \rangle (1 + \langle n_{1k} \rangle)(\omega^2 + \tau'^{-2}_{1k}) c_s^2 N_1 \tau'_{1k}}. \tag{6.57}$$

We have only to find the relaxation time τ' dependence on k. Let us admit that it is defined by the quasi-elastic mechanism of phonon scattering by the density fluctuations. In order to give to explicit expression for the Hamiltonian of this mechanism we shall expand the energy density $\varepsilon(\rho)$ into a series in terms of the power of the small deviations $\delta \rho = \rho - \rho_0$. With a precision of up to terms of the third order of smallness in $\delta \rho$ we have

$$\varepsilon(\rho_0 + \delta \rho) \cong \varepsilon(\rho_0) + \left(\frac{\partial \varepsilon}{\partial \rho_0}\right)_s \delta \rho + \left(\frac{\partial^2 \varepsilon}{\partial \rho_0^2}\right) \frac{\delta \rho^2}{2} + \left(\frac{\partial^3 \varepsilon}{\partial \rho_0^3}\right)_s \frac{\delta \rho^3}{6}. \tag{6.58}$$

Since $\delta \rho = \delta \rho_F + \delta \rho_S$, where $\delta \rho_S$ is the phonon density deviation, $\delta \rho_F$ is fluctuation density deviation from which scattering takes place and the term of interest is proportional to $\delta \rho_F \delta \rho_S^2$, then taking into account that

$$\begin{cases} \left(\dfrac{\partial \varepsilon}{\partial \rho_0}\right)_s = \dfrac{P}{\rho_0^2}, & \left(\dfrac{\partial^2 \varepsilon}{\partial \rho_0^2}\right)_s = \dfrac{c_s^2(m)}{\rho_0} - \dfrac{P}{\rho_0^2}, \\[3mm] \left(\dfrac{\partial^3 \varepsilon}{\partial \rho_0^3}\right)_s = \dfrac{2P}{\rho_0^3} - \dfrac{2c_s^2(m)}{\rho_0^2} + 2\left(\dfrac{c_s(m)}{\rho_0}\right)\left(\dfrac{\partial c_s(m)}{\partial \rho_0}\right)_s \end{cases}$$

(remember that S is the entropy, P is the "pressure", ρ_0 is the equilibrium density value, which is $\rho_0 = (1-m)\rho_1 + m\rho_2$, and $\delta\rho_S = \rho_0 \nabla u$) we find from (6.58) that the interaction Hamiltonian is

$$H = \int \varepsilon(\rho)\mathrm{d}^3 x = \int \delta\rho_F (\mathrm{div}\boldsymbol{u})^2 Q(P,T)\mathrm{d}^3 x \,, \qquad (6.59)$$

where the function

$$Q(P,T) = \frac{2P}{\rho_0} - 2c_s^2(m) + 2\rho_0 \qquad (6.60)$$

$$\times \left[\left(\frac{c_s(m)}{\rho_0} \right) \left(\frac{\partial c_s(m)}{\partial \rho_0} \right)_T + \left(\frac{T^2}{\rho_0^2 c_s} \right) \left(\frac{\partial c_s(m)}{\partial T} \right)_\rho \left(\frac{\partial P}{\partial T} \right)_\rho \right] .$$

Note, that in (6.60) we have passed to variables ρ and T, which are more convenient in terms of analysis.

At last, since the displacement is

$$\boldsymbol{u} = \sum_{k,j} \boldsymbol{e}_j \left[\frac{\hbar}{2\rho_0 V_1 c_s(m)k} \right]^{1/2} [b_k^+(t) + b_{-k}(t)]\mathrm{e}^{\mathrm{i}kx} \,,$$

where $V_1 = V - V_2$ and V_2 is the pore volume, then substituting the given relation into (6.60) we get the final expression for the interaction hamiltonian sought for

$$H = \sum_{\{k\}} \psi(\boldsymbol{k},\boldsymbol{k}',\boldsymbol{k}'')\delta\rho_k(b_{k'}^+ b_{k''} + b_{k'} b_{k''}^+)\Delta(\boldsymbol{k}' - \boldsymbol{k}'' \pm \boldsymbol{k}) \,, \qquad (6.61)$$

where the amplitude of the process is

$$\psi(\boldsymbol{k},\boldsymbol{k}',\boldsymbol{k}'') = - \frac{Q(P,T)(\boldsymbol{e}_j\boldsymbol{k}')(\boldsymbol{e}_j\boldsymbol{k}'')}{2N_1\rho_0(m)c_s(m)(k'k'')^{1/2}} \,. \qquad (6.62)$$

If now according to the known rule (e.g., see monograph [94]) we calculate the probability of a phonon with wave vector k being scattered by density fluctuations, we find

$$\tau_{1k}^{-1} = \left[\frac{Q^2(P,T)k^2 V_1}{120\pi\rho_0^2 c_s^3(m)N_1} \right] \int_0^{2k} p\langle\delta\rho_p^2\rangle\mathrm{d}p \,. \qquad (6.63)$$

At last, in order to calculate the integral involved we should define the distribution function of the density fluctuations. Let us assume that

$$\langle\delta\rho_p^2\rangle = \frac{\rho_0^2 T}{\hbar c_s p} \,.$$

Then finally

$$\tau_{1k}^{-1} = \frac{Q^2(P,T)Tk^3 V_1}{60\pi c_s^4(m)N_1} \,. \qquad (6.64)$$

Thus, substituting (6.64) into (6.57) and taking into account (6.43) we get the sought-for sound energy losses in the solid matrix for $\omega \ll 1/\tau'_{1k}$:

$$\gamma_1(\omega) = \left[\frac{15g^2\hbar(1-m)\omega^2 N_1}{2\pi\rho_0 c_s^2 Q^2(P,T)} \right] \ln\left(\frac{k_{max}}{k_{min}}\right) , \qquad (6.65)$$

where k_{min} and k_{max} are the limiting values of the wave vectors which a phonon may have.

Now let us calculate the contribution to the attenuation $\gamma(\omega)$ which may take place due to the sound wave scattering effect in free volumes. Here two variants are possible:

1. The sound wavelength λ_{sound} is large compared to the maximum linear pore dimension R_{max};
2. λ_{sound} is small compared to the minimum pore dimension R_{min}.

In the first case inhomogeneous sound oscillations in the pores are insignificant and we may neglect the inhomogeneous term on the left-hand side of the kinetic equation (the gradient in the distribution function will vanish), but in the second case it should be taken into account.

The more realistic situation, of course, relates to the first case, which we shall dwell upon. Indeed, if the condition $\lambda_{sound} \gg R_{max}$ is valid, then for R_{max} of order of 10^{-5} cm ω complies with the whole real range of sound frequencies. Let us now solve (6.48) for δn_{2p} in this case. We have

$$\frac{\partial n_{2p}}{\partial t} = -\frac{\delta n_{2p}}{\tau'_{2p}} . \qquad (6.66)$$

In solving this equation we shall follow the paper [116] (see also [117]) in which the sound attenuation coefficient in a Maxwell gas in an external alternating acoustic field was calculated. Taking, therefore, the molecular energy in the interaction with the sound wave as equal to

$$\varepsilon(p,t) = \varepsilon_0(p) + \frac{\boldsymbol{p}\mathrm{d}\boldsymbol{u}}{\mathrm{d}t} , \qquad (6.67)$$

where $\varepsilon_0(p) = p^2/2M$, then for $\delta n_{2p} = n_{2p} - \langle n_{2p} \rangle$ from (5.66) we find

$$\frac{\partial \delta n_{2p}}{\partial t} + \left(\frac{\partial\langle n_{2p}\rangle}{\partial \varepsilon_0}\right)\left(\frac{\mathrm{d}\varepsilon_0}{\mathrm{d}t}\right) = -\frac{\delta n_{2p}}{\tau'_{2p}} . \qquad (6.68)$$

As in the case of phonons, we look for the solution of (6.68) in the form of a plane wave

$$\delta n_{2p}(t) = \delta n_{2p}(0)\mathrm{e}^{-i\omega t} .$$

Therefore, we immediately obtain

$$\delta n_{2p}(0) = \frac{\boldsymbol{p}\boldsymbol{u}_0\omega^2\partial\langle n_{2p}\rangle/\partial\varepsilon_0}{-i\omega + 1/\tau'_{2p}} . \qquad (6.69)$$

Substituting now the obtained solution into (6.47) we have

$$\mathrm{Re}\left(\frac{\langle \partial E_2'/\partial t \langle}{2\rho_0\omega^2 u_0^2 V}\right) = mZ^{-1}N_2^{-1}$$

$$\times \sum_i v_{2i} \int \frac{d^3 p p^2 \omega^2 (\tau_{2p}'^{-2} - \omega^2)\exp(-\varepsilon_0/T)}{\rho_0 TV(2\pi\hbar)^3 \langle n_{2p}\rangle \tau_{2p}'(\omega^2 + 1/\tau_{2p}')^2} \cdot \quad (6.70)$$

Taking $m = \sum v_{2i}/V$ and calculating the integral with respect to momentum at $\omega \ll 1/\tau_{2p}'$ we find that

$$\gamma_2(\omega) = \frac{4\pi m^2 \omega^2 \langle \tau_{2p}'\rangle (2MT)^{5/2}}{3T\rho_0(2\pi\hbar)^3}. \quad (6.71)$$

Adding now relations (6.71) and (6.65) we finally obtain the formula defining the sought-for coefficient of the sound wave energy losses for scattering in a porous structure:

$$\gamma(\omega) = \gamma_1(\omega) + \gamma_2(\omega) = \left(\frac{\omega^2}{\rho_0}\right) \quad (6.72)$$

$$\times \left\{ \frac{15g^2\hbar(1-m)\omega^2 N_1 \ln(k_{\max}/k_{\min})}{2\pi c_s(m)^2 Q^2(P,T)V_1} + \frac{4\pi m^2 \omega^2 \langle \tau_{2p}'\rangle (2MT)^{5/2}}{3T(2\pi\hbar)^3} \right\}.$$

For low concentrations of pores ($m \ll 1$) and low sound velocities $c_s \ll (P/\rho)^{1/2}$ we find

$$\gamma(\omega) \cong \frac{15g^2\hbar\rho_0 c_s^2\omega^2 N_1 \ln(k_{\max}/k_{\min})}{2\pi P^2 V_1}. \quad (6.73)$$

As can be seen from (6.73) for $Q(P,T)$ tending to zero, which is reached for certain relations between the "pressure" P, on the one hand, and the temperature velocity on the other hand, an abrupt increase of the absorption takes place, and the coefficient $\gamma(\omega)$ tends to infinity. Since the absorption cross-section $\sigma(\omega)$ is proportional to $\gamma(\omega)$, it also grows abruptly (see Fig. 6.4) and in this case the whole two-phase system is heated by the external sound source. This, in turn, allows us to conclude that a specific thermal effect takes place due to which the structure of the porous dielectric heats up.

Summarizing, we note that the above theory will be valid if the following rather strict conditions imposed on the wavelength of the external sound are satisfied:

$$\lambda_s \gg l_{fl} > 2r_{fl}. \quad (6.74)$$

Or, since $k = 2\pi/\lambda$, we obtain the following limitation on the sound frequency:

$$\omega \ll \frac{c_s}{l_{fl}} < \frac{c_s}{2r_{fl}}. \quad (6.75)$$

Taking, for example, $l_{fl} = 10^{-4}$ cm and $c_s = 2 \times 10^5$ cm/s, we obtain: $\omega \ll 2 \times 10^9$ Hz. For the opposite inequality, $\omega \gg c_s/2r_{fl}$, in the relevant

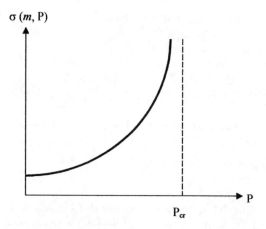

Fig. 6.4. Pressure dependence of the sound absorption cross-section

calculations it is necessary to take into account the inhomogeneity of the
fibril structure. In the kinetic equation describing the evolution behavior of
the distribution function in phase space, in calculating the correction to the
equilibrium distribution function the inhomogeneous term connected with
differentiation with respect to the coordinates should remain. This will not
greatly complicate the calculations, though they will become more awkward.

6.3 The Fluctuation Absorption Mechanism of the Electromagnetic Field

Let us discuss a highly porous substance. Let the porosity m satisfy the
inequality $1 - m \ll 1$. This inequality, by the way, includes the implicit
condition that the structure density ρ is lower than the fibril density ρ_{fl}.
Highly porous substances are not ideal in terms of their practical application
(low mechanical strength, weak electric breakdown strength, and also, if it
concerns, say, cellulose or its derivatives, an inclination to water receptivity).
However, the application of this class of structures in industry allows us to
cope effectively with, for example, various harmful types of electromagnetic
radiation that are always plentiful in air and interfere with some sensitive
radio equipments. In connection with possible applications in technology a
number of relevant questions of theoretical content arises. The first one is the
precise estimation of electromagnetic radiation losses in a porous dielectric
and finding some model regularities applicable to the large class of such
structures. Analysis of the absorption of electromagnetic (EM) radiation will
allow us to predict and find out a number of new specific properties, and
in addition, the theoretical aspect is very important epistemologically and
methodologically.

We remember that in the radio-frequency range the electric losses are known [118] to be estimated by means of the tangent delta $(tg\delta)$; the measurement of the dielectric permeability ε allows us to do so. Indeed, if the alternating field $E(t)$ is applied to a sample, the energy absorbed by it is $E_{abs} = \varepsilon''(\omega)\langle E^2(t)\rangle V/4\pi$, where the angular brackets mean time averaging, $\varepsilon''(\omega)$ is the imaginary part of the dielectric permeability and V is the volume. Such experiments were carried out not only with solid structures but also with various types of cellulose, its derivatives and various sorts of paper, described in part in monographs [1, 23, 93] and in experiments [119–122]. In the EM range, which corresponds to frequency values from 10^{12} up to 10^{15} Hz and to radiation wavelengths from 10^{-5} up to 10^{-2} cm, for fibril structures in which the dimensions of the inhomogeneties (fibrils) are approximately 10^{-6}–10^{-3} cm such radiation must be effectively absorbed. To clarify this we shall refer to the nonstationary relaxation equation describing the fibril distribution over its dimensions, for this approach will enable us to calculate $\varepsilon''(\omega)$ and hence estimate E_{abs}.

In the present problem we shall not include in our analysis the model of f-points (Chap. 2) but shall assume a chaotic ensemble of fibrils placed at arbitrary distances b from each other. Let

$$b \gg l_{fl} > 2r_{fl}. \tag{6.76}$$

The above condition is the criterion for a structure rarefaction, though another case will be analyzed when b becomes of the order l_{fl}.

The case (6.76) is interesting because it allows us to calculate the radiation losses in the gaseous phase purely macroscopically, using only the formula for Raleigh's attenuation [11], and to consider scattering by each fibril independently. Further, let the gaseous phase itself be rather dense so that the EM wavelength satisfies the inequality:

$$\lambda \gg l_M, \tag{6.77}$$

where l_M is the molecular free path in gas. And, at last, the condition to be observed is

$$2r_{fl} < l_{fl} \ll \lambda \ll b. \tag{6.78}$$

For convenience of presentation we shall divide the present section into two parts: in the first part we shall demonstrate the method of calculation for $\alpha''(\omega)$, and in the second part for the absorption cross-section of EM radiation by a fibril ensemble.

To review fluctuation theory we recommend the reader to see Appendix C.

6.3.1 Calculation of $\alpha''(\omega)$

Let the sample under investigation be put in a stationary electric field E_0 whose direction will be taken as the z axis. During the polarization of each fibril as a result of the striction effect [37] it will grow (or compress) in size along this direction and will become a dipole with polarization P. If now we apply along the z axis an alternating and inhomogeneous electric field $E'_z = E'_0 e^{iqx - i\omega t}$, where the frequency is $\omega = cq/\sqrt{\varepsilon}$, c is the speed of light in a vacuum, ε is the dielectric permeability, the wave vector is $q = 2\pi/\lambda$, and λ is the wavelength, due to its small dimensions some frequency in general not equal to ω), and this will be that mechanism which contributes to the absorption of EM a fibril on exposure to alternating field will begin "to breath" oscillate with field energy. The fact that we deal only with the electric field and do not consider the magnetic field is explained by the possibility to apply electrostatic in the range of wavelength in question (condition 6.78).

In order to calculate the z-component of the dielectric susceptibility tensor, which we shall denote simply as $\alpha''(\omega)$, conditioned by this absorption mechanism we shall use a phenomenological relaxation equation which adequately describes the nonstationary change of density of the function of the fibril distribution over dimensions v:

$$\frac{\partial f_v}{\partial t} = -\frac{\delta f_v}{\tau_v}, \tag{6.79}$$

where the small change of density of the distribution is $\delta f_v = f_v - \langle f_v \rangle$. The time τ_v we shall calculate, and the equilibrium value of the function $\langle f_v \rangle$ will be discussed later.

Since we have the relation (6.78) in the volume of fibrils, the field $E'(x, t)$ may be regarded as homogeneous, and hence in the formulae given below we take $q = 0$.

So, let the current volume of the fibrils in the periodic field be

$$v^* = v - d(E_0 + E'(t)) \frac{\cos \theta}{P_{fl}}, \tag{6.80}$$

where θ is the angle between the directions of the dipole moment and the external field and P_{fl} is the "pressure" in the fibrils. Its part may be played, for example, by the Young's modulus [21] or the density of exchange interaction ($P_{fl} = J_{ex}/\langle a \rangle^3$, J_{ex} is the exchange energy, $\langle a \rangle$ is the average interatomic distance $\langle a \rangle = v_{fl}/n_{fl}$, v_{fl} is the fibril volume, and n_{fl} is the entire number of atoms in its volume). Substituting into the left-hand side of (6.79) the distribution density in the form $f_v = \langle f_v \rangle + \delta f_v$, then for the correction δf_v taking account of (6.80) we get the equation sought for:

$$\left(\frac{\partial \langle f_v \rangle}{\partial v} \right) \left(\frac{dv^*}{dt} \right) + \frac{\partial \delta f_v}{\partial t} = \frac{\delta f_v}{\tau_v}. \tag{6.81}$$

We shall look for the oscillating solution of the above equation of the form:

$$\delta f_v(t) = \delta f_v(0) e^{-i\omega t}.$$

Therefore, including (6.80) we obtain

$$\delta f_v(0) = \left(\frac{\partial \langle f_v \rangle}{\partial v} \right) \left[\frac{(dE'(t)/dt)}{P_{fl}} \right] i\omega\tau_v \frac{\cos \theta}{(1 + i\omega\tau_v)}. \tag{6.82}$$

Since the average value of the dipole moment is

$$\langle d_z(t) \rangle = \langle d + \alpha(\omega) v E'(t) \rangle ,\tag{6.83}$$

where $\langle d \rangle$ is its average thermodynamic equilibrium value, for the determination of the functional dependence of $\alpha(\omega)$ we should by means of $f_0 = \langle f_v \rangle + \delta f_v$ calculate $\langle d_z(t) \rangle$. For it we have

$$\langle d_z(t) \rangle = d \langle \int_0^\infty \cos\theta f_v(t) dv \rangle ,\tag{6.84}$$

where the angle brackets imply averaging over angles θ:

$$\langle (\ldots) \rangle = \frac{\int (\ldots) \exp(-dE_0 \cos\theta/T) \sin\theta d\theta}{\int \exp(-dE_0 \cos\theta/T) \sin\theta d\theta} .\tag{6.85}$$

Thus, comparing relations (6.83) and (6.84) and using the solution (6.82) we find that

$$\alpha(\omega) = \left(\frac{d^2}{P_{fl}}\right) \langle v \rangle^{-1} \int_0^\infty \left(\frac{\partial \langle f_v \rangle}{\partial v}\right) i\omega\tau_v \langle \cos^2\theta \rangle \frac{dv}{(1 + i\omega\tau_v)} ,\tag{6.86}$$

where $\langle v \rangle = \int v f(v) dv$.

Calculating further by means of (6.85) the average thermodynamic value of the squared cosine we obtain for the susceptibility:

$$\alpha(\omega) = \left(\frac{d^2}{P_{fl}}\right) \langle v \rangle^{-1} \varphi(\beta) \int_0^\infty \left(\frac{\partial \langle f_v \rangle}{\partial v}\right) i\omega\tau_v dv (1 + i\omega\tau_v) ,\tag{6.87}$$

where

$$\begin{cases} \varphi(\beta) = 1 + 2\beta^{-2} - 2\beta^{-1} cth\beta , \\ \beta = \dfrac{dE_0}{T} . \end{cases}\tag{6.88}$$

The asymptotic value of the function $\varphi(\beta)$ is

$$\varphi(\beta) = \begin{cases} 1 - \dfrac{2}{\beta} & \text{at } \beta \gg 1 , \\ \dfrac{1}{3} + \dfrac{13\beta^2}{180} & \text{at } \beta \ll 1 . \end{cases}\tag{6.89}$$

In order to calculate the remaining integral we have to know two values which depend on the fibril volume: the density of the distribution function and the relaxation time τ_v.

Let us consider two particular cases as follows.

1. Let the fibril distribution in dimensions be subject to the exponential law $\langle f_v \rangle = v_0^{-1} e^{-v/v_0}$, where v_0 is the mathematical expectation or the magnitude of the volume average over the ensemble. The inverse time is

$$\frac{1}{\tau_v} = \left(\frac{v_0}{v}\right)\left(\frac{1}{\tau_0}\right) \tag{6.90}$$

(for details of the τ_v calculation see below).

In this case the integral calculation (6.87) leads to the following value for the imaginary part of the susceptibility:

$$\alpha''(\omega) = \left(\frac{d^2\varphi(\beta)}{v_0^2 P_{fl}\omega\tau_0}\right)\left\{\cos(\omega\tau_0)^{-1}ci(\omega\tau_0)^{-1} - \sin(\omega\tau_0)^{-1}si(\omega\tau_0)^{-1}\right\},$$

$$\tag{6.91}$$

where the sine and cosine integrals are given by

$$\begin{cases} six = -\int\limits_x^\infty \sin t\,\dfrac{dt}{t}\,, \\[2mm] cix = -\int\limits_x^\infty \cos t\,\dfrac{dt}{t}\,. \end{cases} \tag{6.92}$$

The asymptotic values of the relation (6.91) with (6.92) are as follows:

$$\alpha''(\omega) = \left[\frac{d^2\varphi(\beta)}{P_{fl}v_0^2}\right]\begin{cases} \dfrac{C - \ln(\omega\tau_0)}{(\omega\tau_0)} & \text{at } \omega\tau_0 \gg 1\,, \\[3mm] 2\sin\left(\dfrac{2}{\omega\tau_0}\right) & \text{at } \omega\tau_0 \ll 1\,, \end{cases} \tag{6.93}$$

where C is Euler's constant. For information $C = 0.5772$.

From the obtained relation we see that in the region of large frequencies the imaginary part of the susceptibility α'' behaves as $(C - \ln\omega\tau_0)/\omega\tau_0)$, and at small frequencies ω strongly oscillating fluctuations take place, which are visually illustrated in the schematic Fig. 6.5. The nonstandard behavior of α'' leads to the conclusion that such absorption intensity is inherent only for low frequencies.

The reason for the obtained susceptibility behavior is as follows. During time τ_0^* (much smaller than the period of change of the field $2\pi/\omega$) a fibril, being a "heavy" system, "jumps through" the equilibrium state by intertia, and in order to return the dipole moment instantly, in time $\delta t \ll \tau_0^*$, reverses its direction, tending to return to its equilibrium state. But this state is also "jumped through" (again due to the great intertia of the fibril structure), and the situation periodically repeats on the background of the large period $2\pi/\omega$. The above mechanism may be conventionally called the fluctuation mechanism of absorption.

2. Let now the density of the distribution function have the Lorentz form:

$$\langle f_v \rangle = \frac{\pi^{-1}v_0}{(v_2 + v_0^2)}\,. \tag{6.94}$$

The relaxation time we shall, naturally, leave unchanged (see (6.90)). Then substitution of (6.94) and time τ_v from (6.87) after simple calculations yields the dependence:

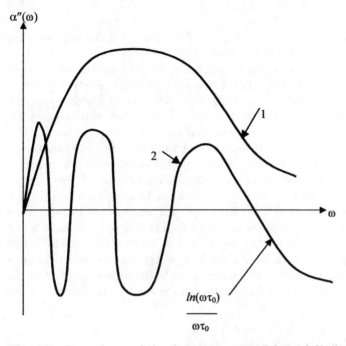

Fig. 6.5. Dependence of the dielectric susceptibility of fibril structures on the field frequency for a Boltzmann distribution of the fibrils over their dimensions. Curve 1 is the Debye dependence of the susceptibility on the frequency; curve 2 is the oscillatory law of the $\alpha'' \langle \omega \rangle$ variation.

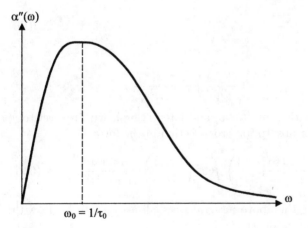

Fig. 6.6. Behavior of α'' for Lorentz law of fibril distribution over its dimensions

$$\alpha''(\omega) = \frac{d^2\varphi(\beta)\omega\tau_0}{2P_{fl}v_0^2}(1+\omega\tau_0)^2 .$$ (6.95)

The plot of $\alpha''(\omega)$ is shown in Fig. 6.6. Note that with the increase of v_0 the peak of the function (6.95) abruptly shifts to zero.

The described approach using the phenomenological kinetic equation not only allows us to calculate $\alpha(\omega)$, but also to predict the rather nonstandard (having, for example, an oscillatory character) absorption capacities of a porous dielectric in an electromagnetic field. This, of course, greatly depends on the fibril volume distribution and on the time τ_v. We also emphasize that the given calculations will be valid only if the distribution of the fibrils in a volume is rather scarce and the electromagnetic radiation wavelength exceeds the maximum linear fibril dimension. If their concentration increases the Van der Waals interaction between them will become significant (see Sect. 3.4). The latter, in turn, means that in calculating $\alpha(\omega)$ we cannot use the formula (6.80) in which this interaction is not taken into account, but should use the general relation for v^* including all the "exchange" components. As a rule, however, in most practical cases the fibrils occupy smaller a space compared to the region occupied by the gaseous phase, which is a justification argument for the theory derived above.

6.3.2 Absorption Cross-Section $\sigma(\omega)$

Let us now calculate the absorption cross-section of the electromagnetic stream by a fibril ensemble. For this purpose we shall use the known dependence $\sigma(\omega)$ (see [123, 124]), which in our case may be written as

$$\sigma(\omega) = 4\pi v^2 \omega \alpha''(\omega, v)\frac{\varepsilon^{1/2}}{c} ,$$ (6.96)

where according to (6.87)

$$\alpha''(\omega, v) = \frac{\alpha''(\omega)}{\langle v \rangle} = -\left(\frac{d^2}{P_{fl}}\right)\left(\frac{\partial\langle f_v \rangle}{\partial v}\right)\frac{\omega\tau_v\varphi(\beta)}{(1+\omega^2\tau_v^2)} .$$

Substituting this relation into (6.96) and integrating it with respect to v, we obtain the sought-for absorption cross-section in the form:

$$\sigma(\omega) = -\left(\frac{4\pi\omega\varepsilon^{1/2}d^2\varphi(\beta)}{P_{fl}c}\langle v \rangle^{-1}\right)\int_0^\infty v^2\left(\frac{\partial\langle f_v \rangle}{\partial v}\right)\frac{\omega\tau_v dv}{(1+\omega^2\tau_v^2)} .$$ (6.97)

In order to calculate the above integral we shall assume, to begin with, the exponential distribution density:

$$\langle f_v \rangle = v_0^{-1}\exp\left\{-\frac{v}{v_0}\right\} .$$

As a result, the absorption cross-section is

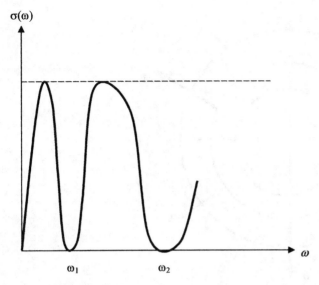

Fig. 6.7. Oscillatory behavior of the cross-section for the absorption of electromagnetic radiation by a fibril structure

$$\sigma(\omega) = \left(\frac{4\pi\varepsilon^{1/2}d^2\varphi(\beta)}{P_{fl}cv_0\tau_0} \right) \tag{6.98}$$
$$\times \left\{ 1 + (\omega\tau_0)^{-1}[\cos(\omega\tau_0)^{-1}ci(\omega\tau_0)^{-1} - \sin(\omega\tau_0)^{-1}si(\omega\tau_0)^{-1}] \right\} .$$

The asymptotic behavior of $\sigma(\omega)$ is

$$\sigma(\omega) = \left[\frac{4\pi\varepsilon^{1/2}d^2\varphi(\beta)}{P_{fl}cv_0\tau_0} \right] \tag{6.99}$$
$$\times \begin{cases} 1 - (\omega\tau_0)^{-1}[0.5\pi + (\omega\tau_0)^{-1}\ln(\omega\tau_0)] & \text{at } \omega\tau_0 \gg 1, \\ 1 - \cos[2(\omega\tau_0)^{-1}] & \text{at } \omega\tau_0 \ll 1. \end{cases}$$

It is seen that in the low-frequency range the absorption cross-section $\sigma(\omega)$, as well as α'', has an abruptly oscillating character. Since the period on the frequency scale is $\Delta\omega = 2\pi\omega^2\tau_0$, it becomes clear that for very small frequencies the period is small, and when ω increases the period of oscillations increases as the square of the frequency. Most visually it is shown in Fig. 6.7.

Let now the dimension distribution of the fibrils be the Lorentz distribution with probability density (6.94). In this case the calculation of $\sigma(\omega)$ by means of the formula (6.97) gives the following relation:

$$\sigma(\omega) = \left[\frac{4\pi\varepsilon^{1/2}d^2\varphi(\beta)}{P_{fl}cv_0\tau_0} \right] B(x), \tag{6.100}$$

where $B(x) = x^2(x^2 - 1 - 2\ln x)/(x^2 - 1)$ and $x = (\omega\tau_0)^{-1}$.

As we see, at large dimension spread of fibrils and low frequencies ($\omega\tau_0 \ll 1$ the curve $\sigma(\omega)$ tends to increase abruptly.

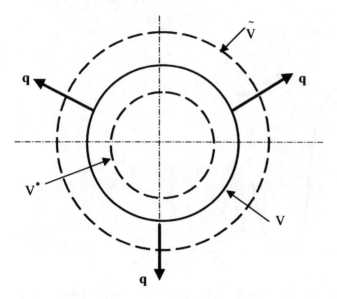

Fig. 6.8. Fluctuation change of fibril dimensions. "Intrinsic" frequency is $\omega_n = (\pi n \tau_0)^{-1}$ at $n \gg 1$

Now, as we promised above, we shall calculate the time τ_v. Imagine one isolated fibril and assume that with temperature decrease its dimension will decrease too (see Fig. 6.8), but that the internal energy does not change! Then

$$\frac{d\varepsilon}{dt} = T\frac{dS}{dt} - P_{fl}\frac{dv}{dt} = 0\,. \tag{6.101}$$

In a cylindrical system of coordinates and on condition that the fibril radius r_F changes very slightly, Eq. (6.101) may be rewritten in the form:

$$\chi\frac{\partial^2 T}{\partial r_0^2} = P_{fl}s_0\frac{dr_0}{dt}\,, \tag{6.102}$$

where χ is the thermal diffusion coefficient of a fibril and s_0 is its surface square.

On the other hand, we have the boundary condition:

$$\kappa\left.\frac{\partial T}{\partial r}\right|_{r=r(t)} = -\alpha(T - T_0)\,, \tag{6.103}$$

where $\kappa = \chi c_{flv}$, c_{flv} is the thermal capacity and α is the coefficient of thermal transfer from the fibrils to the gaseous phase. T_0 is some temperature determined below. Solving (6.103) for the conditions

$$\begin{cases} r_0(t)|_{t=0} = r_0(0)\,, \\ r_0(t)|_{t\Rightarrow\infty} = r^* \end{cases} \tag{6.104}$$

with $r^* \leq r_0(t) \leq r_0(0)$, we obtain

$$T(t) = T_0 - \frac{T(0) - T_0}{\exp\{-\alpha[r^* - r_0(0)]/\kappa\} - 1}$$

$$+ \frac{[T(0) - T_0]\exp\{-\alpha[r_0(t) - r^*]/\kappa\}}{\exp\{-\alpha[r^* - r_0(0)]/\kappa\} - 1}. \tag{6.105}$$

Substituting (6.105) into (6.102) and assuming that $r_0(t)$ is close to r^*, after relevant expansions we find

$$\frac{dr_0}{dt} = -\tau^{*-1}(r_0 - r^*). \tag{6.106}$$

Then, omitting the asterisk from r and passing to the volume v we get the final equation sought for:

$$\frac{dv}{dt} = \tau_v^{-1}(v_0 - v), \tag{6.107}$$

where the sought-for relaxation time which we used in the above text is determined by the formula:

$$\frac{1}{\tau_v} = \frac{v_0\alpha^2[T_0 - T(0)]}{\kappa P_{fl}(v - v^*)} = \left(\frac{v_0\alpha^2}{P_{fl}}\right)\left(\frac{1 - c_p}{c_v}\right)\left(\frac{\partial T}{\partial v}\right)_P, \tag{6.108}$$

where c_p and c_v are the isobaric and isochoric thermal capacities of the fibrils, respectively.

Thus, we have assured ourselves that the relaxation time τ_v does depend on the volume linearly, which justifies the relation (6.90) introduced phenomenologically.

As to the order of magnitude of the frequencies, for which the present consideration is valid, we should estimate the inverse relaxation time τ_v using the obtained formula (6.108). Indeed, we have

$$\tau_v^{-1} = \frac{v_0\alpha^2\Delta T}{\kappa P_{fl}\Delta v} \cong \frac{\alpha^2 T}{\kappa P_{fl}},$$

where the coefficient of thermal transfer from the fibrils to the gaseous phase is $\alpha = 1/\tau s_0$ and the thermal conductivity coefficient is $\kappa = c_v v_T^2 \tau$. Assuming, for example, $T = 300$ K, $P_{fl} = 10^{10}$ erg/cm^3, $v_T = 10^5$ cm/s, $\tau = 10^{-10}$ s, $s_0 = 10^{-8}$ cm^2 and $c_v = 10^{20}$ 1/cm^3, we find $\tau_v^{-1} = 10^4$ Hz. Hence, the desired frequency interval should be from the infrared spectrum band to the ultralow band of the order of 10^4 Hz.

Table 6.1, taken from [65], shows the dependence of the tangent delta on the temperature and frequency of the electromagnetic field. The graphical representation of the functional dependence of tgδ on the temperature for different frequencies is given in Fig. 6.9. Figure 6.10 shows the frequency dependence of dielectric permeability of cellulose at fixed temperature $T = 288$ K, where ε_0 is the static permeability at $\omega = 0$.

Now we briefly formulate the conclusions of the present section:

Table 6.1. Temperature-frequency characteristics of the dielectric permeability ε and the loss angle ($tg\delta$) of cellulose at frequencies of the external field of 1–40 MHz

Temperature T (°C)	Parameter	Frequency (MHz)				
		1	10	20	30	40
15	ε	6.62	6.45	6.28	6.11	5.95
	$10^4 tg\delta$	590	742	727	688	642
25	ε	6.60	6.43	6.26	6.09	5.93
	$10^4 tg\delta$	527	742	785	784	760
35	ε	6.58	6.41	6.24	6.07	5.91
	$10^4 tg\delta$	465	671	782	814	827
45	ε	6.56	6.39	6.22	6.05	5.89
	$10^4 tg\delta$	402	584	713	776	815
55	ε	6.53	6.37	6.20	6.03	5.87
	$10^4 tg\delta$	340	496	600	680	740
65	ε	6.51	6.35	6.17	6.01	585
	$10^4 tg\delta$	286	406	488	562	622
75	ε	6.48	6.32	6.15	5.98	5.83
	$10^4 tg\delta$	240	339	400	449	502
85	ε	6.46	6.29	6.12	5.46	5.80
	$10^4 tg\delta$	205	280	328	367	400
95	ε	6.43	6.27	6.09	5.93	5.77
	$10^4 tg\delta$	174	236	277	302	323

1. Methods of calculation of the absorption coefficient of electromagnetic radiation by highly porous structures have been demonstrated.
2. It has been shown that an oscillatory dependence of the dielectric susceptibility $\alpha''(\omega)$ and absorption cross-section $\sigma(\omega)$ exist.

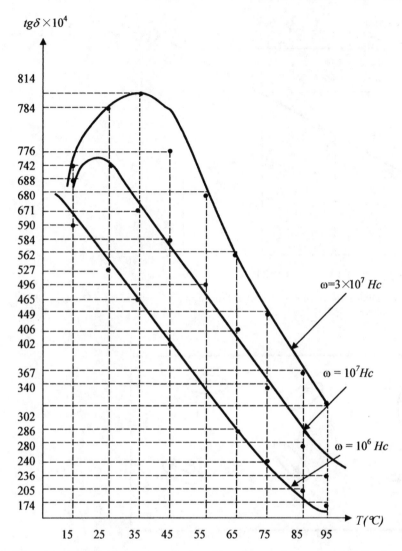

Fig. 6.9. Experimental temperature dependence of the tangent delta for different fixed frequencies for cellulose ([121])

6.4 Dielectric Permeability of Porous Substances

All the calculations given in the present chapter were connected with the study of the influence of external alternating fields on a porous medium. If a porous structure is in a stationary electric field $\boldsymbol{E}_0$, then for the volume distribution characteristic of internal fields the *static* dielectric permeability ε should also be considered. Here the question naturally arises: how will the

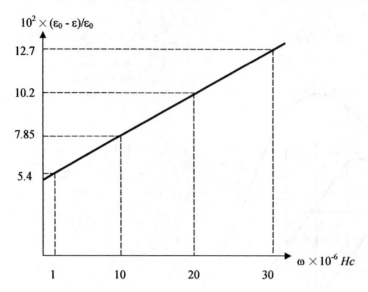

Fig. 6.10. Dielectric permeability of cellulose at $T = 288\,\mathrm{K}$ as a function of the field frequency ω ([121])

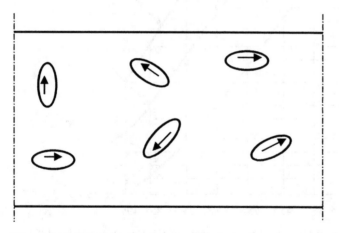

Fig. 6.11. Chaotic arrangement of fibrils in space

static permeability ε of the whole sample depend on the structure porosity m and on the dielectric permeability ε_1 of the fibrils and ε_2 of the gas phase?

Let us assume that the internal structure of the dielectric includes an ensemble of equal (in size) fibrils. Each fibril has a symmetry axis n_j directed along its length l_{fl}, as shown in Fig. 6.11. Let us take a single fibril and introduce a system of coordinates $K'(x', y', z')$ which are connected by this fibril (see Fig. 6.12). In Fig. 6.13 the Euler angles of rotation are depicted (the reader can find a detailed description these angles in any textbook on

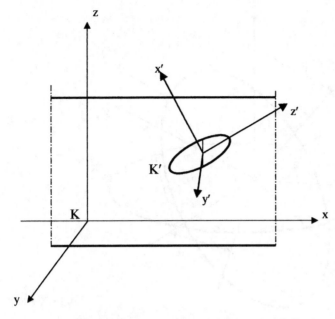

Fig. 6.12. Schematic representation of one isolated fibril and the position of the coordinate axes in system K'

theoretical physics and in particular they are wonderfully described in the monographs of Landau and Lifshitz [55] and Matthews and Hocker [125]), connecting the local system K' of a j-fibril with the global system K. The matrix $a_{\nu\mu}$, connecting the vector components in the system K' with the vector components in the system K is assigned in the form of the linear transformation given by

$$x_\nu = \sum_{\mu=1}^{3} a_{\nu\mu} x'_\mu \,, \tag{6.109}$$

where the indices μ and ν correspond to the x, y, z coordinates and the matrix of rotations given by

$$a_{\nu\mu} = \left\| \begin{matrix} \cos\alpha_j \cos\beta_j \cos\gamma_j - \sin\alpha_j \sin\gamma_j & \sin\alpha_j \cos\beta_j \cos\gamma_j + \cos\alpha_j \sin\gamma_j & -\sin\alpha_j \cos\gamma_j \\ -\cos\alpha_j \cos\beta_j \sin\gamma_j - \sin\alpha_j \sin\gamma_j & -\sin\alpha_j \cos\beta_j \sin\gamma_j + \cos\alpha_j \cos\beta_j \sin\gamma_j & \sin\beta_j \sin\gamma_j \\ \cos\alpha_j \sin\beta_j & \sin\alpha_j \sin\beta_j & \cos\beta_j \end{matrix} \right\| \,. \tag{6.110}$$

The range of variation of the rotation angles is

$$0 \le \alpha_j \le 2\pi \,, \quad 0 \le \beta_j \le \pi \,, \quad 0 \le \gamma_j \le 2\pi \,. \tag{6.111}$$

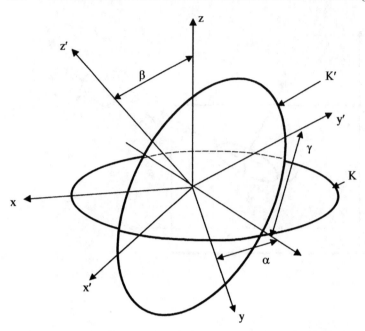

Fig. 6.13. Interrelation of systems K and K' through the Eulerian angles α, β and γ

According to Maxwell's equations, which are satisfied by the internal electric field $E^{(i)}$ in the dielectric, the following two equations may be inferred:

$$\begin{cases} \boldsymbol{D} = \boldsymbol{E}^{(i)} + 4\pi\boldsymbol{P}\,, \\ \boldsymbol{E}^{(i)} = \boldsymbol{E}_0 + 4\pi\boldsymbol{P}(L-N)\,, \end{cases} \tag{6.112}$$

where the electric induction vector is

$$\boldsymbol{D} = \varepsilon\boldsymbol{E}^{(i)}\,. \tag{6.113}$$

The Lorenz field is given by

$$\boldsymbol{E}_L = 4\pi L\boldsymbol{P}\,. \tag{6.114}$$

Here L is the Lorenz factor (in the general case L is a tensor).

The depolarization field connected with the pure sample geometry is

$$\boldsymbol{E}_N = 4\pi N\boldsymbol{P}\,, \tag{6.115}$$

where N is the depolarization tensor and $\boldsymbol{P}$ is the polarization vector (dipole moment of a dielectric unit volume).

Substituting (6.13) into the upper equation of (6.112) and solving the obtained equation relative to $\boldsymbol{P}$ we find

$$4\pi\boldsymbol{P} = (\varepsilon - 1)\boldsymbol{E}^{(i)}\,. \tag{6.116}$$

From the lower equation of (6.112) we find that the internal field is

$$\boldsymbol{E}^{(i)} = \frac{\boldsymbol{E}_0}{1 + (L - N)(\varepsilon - 1)}. \tag{6.117}$$

The induction vector, therefore, is

$$\boldsymbol{D} = \frac{\varepsilon \boldsymbol{E}_0}{1 + (L - N)(\varepsilon - 1)}. \tag{6.118}$$

Then, the internal field of the whole structure is combined from two parts: from components in the gaseous and solid phases. Therefore, taking the porosity into account we may write

$$
\begin{cases}
\boldsymbol{E}^{(i)} = (1 - m)\boldsymbol{E}_1^{(i)} + m\boldsymbol{E}_2^{(i)}, & \tag{6.119} \\[2mm]
\boldsymbol{P} = (1 - m)\boldsymbol{P}_1 + m\boldsymbol{P}_2, & \tag{6.120} \\[2mm]
\boldsymbol{D} = (1 - m)\boldsymbol{D}_1 + m\boldsymbol{D}_2. & \tag{6.121}
\end{cases}
$$

Introducing the permeabilities ε_1 and ε_2 we may rewrite the Eq. (6.121) in another way:

$$\boldsymbol{D} = (1 - m)\varepsilon_1 \boldsymbol{E}_1^{(i)} + m\varepsilon_2 \boldsymbol{E}_2^{(i)}. \tag{6.122}$$

By analogy with (6.117) we find

$$
\begin{cases}
\boldsymbol{E}_1^{(i)} = \dfrac{\boldsymbol{E}_0}{1 + (L_1 - N_1)(\varepsilon_1 - 1)}. & \tag{6.123} \\[4mm]
\boldsymbol{E}_2^{(i)} = \dfrac{\boldsymbol{E}_0}{1 + (L_2 - N_2)(\varepsilon_2 - 1)}. & \tag{6.124}
\end{cases}
$$

Substituting (6.123) and (6.124) into (6.122) and also replacing $\boldsymbol{D}$ by means of (6.118) we find the equation for determination of the sought-for permeability of the porous substance:

$$\frac{\varepsilon}{1 + (L - N)(\varepsilon - 1)} = \frac{(1 - m)\varepsilon_1}{1 + (L_1 - N_1)(\varepsilon_1 - 1)} + \frac{m\varepsilon_2}{1 + (L_2 - N_2)(\varepsilon_2 - 1)}. \tag{6.125}$$

Solving it with respect to ε we obtain

$$\varepsilon = g\frac{1 + N - L}{1 + g(N - L)}, \tag{6.126}$$

where

$$g = \frac{(1 - m)\varepsilon_1}{1 + (L_1 - N_1)(\varepsilon_1 - 1)} + \frac{m\varepsilon_2}{1 + (L_2 - N_2)(\varepsilon_2 - 1)}. \tag{6.127}$$

Let us consider some particular cases of the relation (6.126).

1. For a spherical porous sample we have $N = L$. If for the gaseous phase we take $\varepsilon_2 \cong 1$, then

$$\varepsilon = g = m + \frac{(1 - m)\varepsilon_1}{1 + (L_1 - N_1)(\varepsilon_1 - 1)} . \qquad (6.128)$$

It should be stressed that the coefficients L_1 and N_1 depend on the porosity m! (see monograph [125] and papers [126–136]). If the number of pores is not large, i.e. $m \ll 1$, the difference $L_1 - N_1 \cong K^* m$, where K^* is some constant, and then from (6.128) there follows the relation:

$$\varepsilon = m + (1 - m)\varepsilon_1[1 - K^*(\varepsilon_1 - 1)m]$$
$$\cong \varepsilon_1 - m[\varepsilon_1 + K^*\varepsilon_1(\varepsilon_1 + 1) - 1] . \qquad (6.129)$$

2. If the fibrils are spherical, i.e. $L_1 = N_1 = 1/3$ (again we take $\varepsilon_2 \cong 1$), from (6.127) we find

$$g = (1 - m)\varepsilon_1 + m \qquad (6.130)$$

and

$$\varepsilon \cong \frac{\varepsilon_1 - m(\varepsilon_1 - 1)[(N - L)^2 + 1 + N - L]}{1 + \varepsilon_1(N - L)} . \qquad (6.131)$$

3. If $m = 0$, then $N_1 = N$ and $L_1 = L$, $\varepsilon = \varepsilon_1$. If $m = 1$, then $N_2 = N$, $L_2 = L$ and $\varepsilon = \varepsilon_2$.

The schematic dependence of the dielectric permeability on the structure porosity m is illustrated in Fig. 6.14. We emphasize that such a qualitative

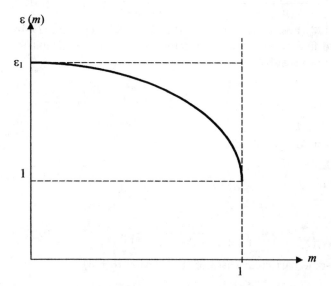

Fig. 6.14. Schematic dependence of the dielectric permeability of a porous substance on the porosity m

behavior of $\varepsilon(m)$ is inherent both for the first case (formula (6.129)) and for the second case (formula (6.131)).

The transition to the fibril ensemble when the number of fibrils N_0 is large is performed by means of the matrices $a_{\nu\mu}$ (6.110) and corresponds to the limit:

$$\lim_{N_0 \Rightarrow \infty} N_0^{-1} \sum(\ldots) = \left(\frac{1}{8\pi^2}\right) \int\limits_0^{2\pi} d\alpha \int\limits_0^\pi \sin\beta d\beta \int\limits_0^{2\pi} d\gamma(\ldots). \qquad (6.132)$$

It should be noted that because of the tensor character of the coefficients $N(N_{1,2})$ and $L(L_{1,2})$ such simple laws of dependence of ε on m as those obtained above will not hold. In this case we have to solve the tensor equation for ε_{ik} and, though it can be solved in the simplest cases, the problem is beyond our current plans.

6.5 Explanation of the Temperature Dependence of the Loss Angle Tangent

The physical picture explaining the relaxation polarization of any dielectric is based on the representation of its internal structure as a medium, for example, the form of a deep potential well with two equilibrium states (the so-called Fröhlich relaxator) of ball-shaped dipoles in a viscous medium, a potential box with electrons or holes, and so on.

The idea of the problem is to find out the relaxation time dependence as a function of the substance parameters and, in particular, of the temperature by means of some model representation.

In dielectrics the relaxation time is understood as the time necessary to achieve structure polarization and it is either the time of rotation of an atom (molecule) dipole moment vector into the direction of the polarizing stationary electric field $\boldsymbol{E}_0$, or the time of separation of the charges into positive (on one side) and negative (on the other side), i.e. the time it takes for the polarizing dipole moment (of polarization vector $\boldsymbol{P}$) to appear.

At the same time it should be remembered that the polarization vector is a dipole moment related to a dielectric unit volume.

The calculation at the above mentioned times is based, as a rule, on a powerful mathematical apparatus and basically is connected with the application of the Boltzmann kinetic equation, the Fokker–Planck diffusion equation, and Smoluchowsky's equation (see [112, 113, 138–141, 143, 151] and also the review [54]).

The derivation in the previous section of the precise equation for the dielectric permeability of a porous dielectric (compare also with formula (2.54) of Chap. 2) which relates ε to the porosity m, enables us to estimate the behavior of tgδ as a function of the frequency of the applied electric field ω, and of the temperature T as well.

We shall consider the pores as spherical inclusions whose radii are held equal and are denoted by R. The dielectric will be chosen in the shape of a plate (or film). In this case from (6.131) it is easy to determine ε. Indeed, taking the electric field as oriented perpendicularly to the sample plane and taking this direction for the z axis and the dielectric plate plane for the x–y axis, for the coefficients L and N we have: $N_{zz} = L_{zz} = 1$, $N_{xx} = N_{yy} = L_{xx} = L_{yy} = 0$. That is, in the given case $N = L$, and as a result the formula (6.131) gives

$$\varepsilon = (1 - m)\varepsilon_1 + m\varepsilon_2 \,. \tag{6.133}$$

Note that since the coefficients N and L are, in general, tensor quantities, in (6.133) the dielectric permeability should be treated as a tensor and therefore the relation (6.133) is the zz-component of tensor ε_{ik}, i.e. $\varepsilon = \varepsilon_{zz} = (1 - m)\varepsilon_1 + m\varepsilon_2$.

Hence it directly follows that the real part of ε, ε', and the imaging part, ε'', are given by

$$\begin{cases} \varepsilon' = (1 - m)\varepsilon_1' + m\varepsilon_2' \,, \\ \varepsilon'' = (1 - m)\varepsilon_1'' + m\varepsilon_2'' \,. \end{cases} \tag{6.134}$$

By definition the tangent "delta" is found as the ratio $\varepsilon''/\varepsilon'$, and therefore, from (6.134) we get

$$\mathrm{tg}\delta = \frac{(1 - m)\varepsilon_1'' + m\varepsilon_2''}{(1 - m)\varepsilon_1' + m\varepsilon_2'} \,. \tag{6.135}$$

In order to find the temperature dependence of $\mathrm{tg}\delta$ on the temperature and frequency we choose the simplest relationship between the real and imaginary parts of the dielectric permeability of each phase in the form of Debye's formulae. That is,

$$\begin{cases} \varepsilon_i' = \varepsilon_{0i} + \dfrac{(\varepsilon_{\infty i} - \varepsilon_{0i})\omega^2\tau_i^2}{1 + \omega^2\tau_i^2} \,, \\[4mm] \varepsilon_i'' = \dfrac{(\varepsilon_{\infty i} - \varepsilon_{0i})\omega\tau_i}{1 + \omega^2\tau_i^2} \,, \end{cases} \tag{6.136}$$

where $i = 1, 2$.

Substituting now (6.136) into (6.135) we get the rather awkward equation

$$\mathrm{tg}\delta = \frac{a_1(1 - m)\omega\tau_1(1 + \omega^2\tau_2^2) + a_2 m\omega\tau_2(1 + \omega^2\tau_1^2)}{b(1 + \omega^2\tau_1^2)(1 + \omega^2\tau_2^2) + a_1(1 - m)\omega^2\tau_1^2(1 + \omega^2\tau_2^2) + a_2 m\omega^2\tau_2^2(1 + \omega^2\tau_1^2)} \,, \tag{6.137}$$

where the new parameters are

$$\begin{cases} a_1 = \varepsilon_{\infty 1} - \varepsilon_{01} \,, \\ a_2 = \varepsilon_{\infty 2} - \varepsilon_{02} \,, \\ b = (1 - m)\varepsilon_{01} + m\varepsilon_{02} \,. \end{cases} \tag{6.138}$$

The above equation allows us to consider the following four possible cases.

1. The frequency of the external alternating electric field is small, i.e. $\omega \ll \tau_1^{-1}, \tau_2^{-1}$. As a result from (6.137) there follows the rather simple equation:

$$\mathrm{tg}\delta = \omega b^{-1}[a_1(1-m)\tau_1 + a_2 m\tau_2].\tag{6.139a}$$

2. If the field frequency is high, i.e. $\omega \gg \tau_1^{-1}, \tau_2^{-1}$, then

$$\mathrm{tg}\delta = \frac{[a_1(1-m)\tau_1^{-1} + a_2 m\tau_2^{-1}]}{\omega[b + (1-m)a_1 + ma_2]}.\tag{6.139b}$$

3. In the intermediate frequency range when $\tau_2^{-1} \ll \omega \ll \tau_1^{-1}$, the general Eq. (5.137) gives

$$\mathrm{tg}\delta = \frac{[a_1(1-m)\omega\tau_1 + a_2 m/\omega\tau_2]}{(b + ma_2)}.\tag{6.139c}$$

4. And finally, for the last intermediate frequency range $\tau_1^{-1} \ll \omega \ll \tau_2^{-1}$:

$$\mathrm{tg}\delta = \frac{[a_1(1-m)/\omega\tau_1 + a_2 m\omega\tau_2]}{(b + (1-m)a_1)}.\tag{6.139e}$$

Thus, we see that for qualitative analysis of the temperature dependence of tgδ we should find the dependence of the relaxation polarization times in the solid and gaseous phases as functions of T.

We shall start with the fibril structure. Assuming that its physical properties may be described in the framework of solid crystal dielectric theory for a wide band gap, we may introduce the time of establishment of charge asymmetry (which is in general, equivalent to the dielectric polarization), which is known to be defined by charge carrier mobility μ (e.g. see monograph [143]). For reference we may show the connection of the conduction σ with the charge carrier mobility, i.e. $\sigma = en_e\mu$, where n_e is the electron concentration and e is the electron charge. According to the review [144] the polarization relaxation time in the solid phase may be presented as $1/\tau_1 = 12T\mu/eS_{fl}$, where $S_{fl} =$ fibril surface area. Taking the fibril to be spherical and using $S_{fl} = 4\pi r_{fl}^2$, we get

$$\frac{1}{\tau_1} = \frac{3T\mu}{\pi e r_{fl}}.\tag{6.140}$$

Note once more that we consider Boltzmann's constant to be equal to one. As to the time of dipole "turning" in gaseous phase "2", as we know from Sect. 4.7, this time may be defined as the time of turning of the dipole moment of every heavy molecule in some viscous medium with viscosity η into the direction of the stationary electric field $\boldsymbol{E}_0$. As was shown in [126], the corresponding relaxation of a magnetic particle in a viscous medium and magnetic field H_0 is defined by the formula $1/\tau = \chi H_0^2/3\pi^2\eta$, where χ is the magnetic susceptibility of a particle. In the case of the dielectric under consideration the role of the product χH_0^2 corresponds to αE_0^2, where α is the molecular polarization coefficient, and the desired "turning" time must be found as

$$\frac{1}{\tau_2} = \frac{\alpha_2 E_0^2}{3\pi^2 \eta_2}, \tag{6.141}$$

where η_2 is the gaseous-phase viscosity in a pore and α_2 is the molecular polarization coefficient in phase "2".

If the stationary electric field is absent ($E_0 = 0$), then the polarization time may be estimated as the time of spontaneous turning of all the heavy gas molecules up to *self-organized* direction of each molecule in external random field acting from the side of the rest ones. This time is determined by a random fluctuation mechanism and it should be estimated in the following way. If we denote the linear dimension of a molecule as a, then in (6.141) it should be stated that $< \alpha_2 E_0^2 a^3 > = T$, where the angular brackets imply simple statistical averaging with respect to the equilibrium canonical Gibbs's distribution. So, in this case the sought-for relaxation time is

$$\frac{1}{\tau_2} = \frac{T}{3}\pi^2 \eta_2 a^3. \tag{6.142}$$

The obtained relation contains the viscosity, which must depend on the temperature. To estimate it we shall use the gas-kinetic approximation and take $\eta_2 = \rho_2 l v_T$, where l is the free path of the gas molecules surrounding the heavy molecule in question and v_T is their average heat velocity. By definition $l = v_T \tau$, where τ is the average time interval between molecular collisions.

Thus, the viscosity is $\eta_2 = \rho_2 v_T^2 \tau$. The average heat velocity of the molecules, as we know, is $v_T^2 = (3T/2M)^{1/2}$, and we have only to estimate the dependence of the gas density from the temperature and relaxation time. For the relaxation time τ from the Boltzmann classic kinetic equation (e.g. see [146]) we get $1/\tau = \sigma n_M v_T$, where σ is the cross-section for molecules scattered by each other and n_M is their concentration in a unit volume (the dimensions of n_M in the SGC system is $[1/cm^3]$). Since σ practically does not depend on the temperature (actually, of course, it does depend but very weakly), it turns out that

$$\frac{1}{\tau} = \sigma n_M v_T \approx a^2 n_M \left(\frac{3T}{2M}\right)^{1/2}, \tag{6.143}$$

where a is the average diameter of a molecule. To estimate temperature dependence of ρ_2 we shall use the ideal gas equation (Clapeyron–Mendeleev equation), according to which $P_2 V_2 = N_2 T$, where P_2 is the pressure and N_2 is the number of molecules in phase "2". Introducing the density as the ratio $\rho_2 = N_2 M / V_2$, we immediately find that $\rho_2 = M P_2 / T$. Thus, the density is $\eta_2 = 3 M^{1/2} P_2 / 2 n_M a^2 T^{1/2}$. Substituting the equation obtained for viscosity into (6.143), we ultimately get the equation for the universal time of polarization relaxation in the gaseous phase:

$$\frac{1}{\tau_2} = \frac{2n_M T^{3/2}}{9\pi^2 a M^{1/2} P_2}.\tag{6.144}$$

So, now we possess all the necessary parametrical relations given by (6.141) and (6.144), which are quite sufficient for adequate explanation of the functional dependence of tgδ on temperature.

If we substitute (6.141) and (6.144) into the general Eq. (6.137), we may get a rather awkward and quite unperceivable for eyes equation. Reluctant though we are, we shall have to give it in explicit form and, at least qualitatively, show the shape of the correct temperature dependence illustrated by the experimental points in Fig. 6.9. For this we shall first slightly simplify the recording of the inverse relaxation times in (6.141) and (6.144) by presenting them as

$$\begin{cases} \dfrac{1}{\tau_1} = B_1 T, \\[2mm] \dfrac{1}{\tau_2} = B_2 T^{3/2}, \end{cases}\tag{6.145}$$

where the new parameters are $B_1 = 3\mu/\pi e r_{fl}^2$ and $B_2 = 2n_M/9\pi^2 a M^{1/2} P_2$.

Substituting now (6.145) into (6.137) and we find

$$tg\delta = \frac{(1-m)a_1 B_1 T\omega(B_2^2 T^3 + \omega^2) + ma_2 B_2 T^{3/2}\omega(B_1^2 T^2 + \omega^2)}{b(B_1^2 T^2 + \omega^2)(B_2^2 T^3 + \omega^2) + (1-m)a_1\omega^2(B_2^2 T^3 + \omega^2) + ma_2\omega^2(B_1^2 T^2 + \omega^2)}.\tag{6.146}$$

From the above equation (now not so awkward) we can draw conclusions about the existence of a temperature maximum, shifting with higher frequency ω to higher temperatures. Let us estimate the qualitative behavior of tgδ in more detail and consider four possible temperature ranges as follows.

1. For $T \ll \omega/B_1, (\omega/B_2)^{2/3}$, from (5.146) it follows that

$$tg\delta = \frac{(1-m)a_1 B_1 T}{[b + (1-m)a_1 + ma_2]};\tag{6.147a}$$

2. For $\omega/B_1 \ll T \ll (\omega/B_2)^{2/3}$, we obtain

$$tg\delta = \frac{[(1-m)a_1\omega^2 + ma_2 B_1 B_2 T^{5/2}]T}{[B_1 T^2(b + ma_2) + B_1^{-1}a_1(1-m)\omega^2]\omega};\tag{6.147b}$$

3. For $(\omega/B_2)^{2/3} \ll T \ll \omega/B_1$, we obtain

$$tg\delta = \frac{\omega T B_2[(1-m)a_1 B_1 B_2 T^3 + ma_2\omega^2 T^{1/2}]}{bB_2 T^3 + (1-m)a_1 B_2^2\omega^2 T^3 + ma_2\omega^4};\tag{6.147c}$$

4. For $T \gg \omega/B_1, (\omega/B_2)^{2/3}$, we obtain

$$tg\delta = \frac{(1-m)a_1\omega}{bB_1 T}.\tag{6.147d}$$

Thus, in the region of relatively low temperatures, as we can see from (6.147a), $tg\delta$ depends linearly on T, and in the region of high temperatures from (6.147d) there follows a hyperbolic dependence, i.e. $tg\delta \sim 1/T$. Here our attention is at once caught by the fact that both limiting temperature "tails" are completely defined by the relaxation in the fibril (solid phase) structure. In intermediate ranges (formulae (6.147b) and (6.147c)) the situation is more bizarre and qualitatively the temperature behavior of the tangent "delta" is defined by the polarization in both phases.

In fact, from (6.147a) and (6.147d), the conclusion follows about the "correct" qualitative behavior of $tg\delta$ and enables us to explain the existence of a maximum shifting for higher frequencies of the external field roughly according to a linear law, which is illustrated by Fig. 6.9, which is based on experimental data [121].

Let us now pay attention to one more thing. Let the frequency of the external alternating electric field be ω_1. Then in order to describe the process of field energy absorption in a dielectric, an effective relaxation mechanism should be found, giving a relaxation time τ^1 which would be comparable with the field period $T_1 = 2\pi/\omega$. For frequencies, say, ω_2 much higher than ω_1 we have to find such a dissipative mechanism whose characteristic time will be denoted by τ^2. This reasoning may be continued further but the principle is clear: the product $\omega_i\tau^i$, where $i = 1, 2, \ldots, n$ and n is the number of discrete changes of the external field frequency, i.e. $\omega_1\tau^1 \sim \omega_2\tau^2 \sim \ldots \sim \omega_n\tau^n$, should remain "invariant". Thus, the following conclusion may be formulated: the higher the frequency of affecting, the faster the relaxation mechanisms that should be looked for. The idea is clear, but to find and, moreover, to describe physical mechanisms which begin to act at different frequencies is a rather complicated task and it depends on correct modeling of the physical structure of the substance under consideration and on the correct understanding of the physical essence of the relaxation.

For frequencies ω of the order of $10^6 - 10^7$ Hz on which the experiment was carried out [121], the times $1/\tau_1$ and $1/\tau_2$ correspond to what has just been stated.

Indeed, let us estimate the times $\tau_1 \equiv \tau^1$ and $\tau_2 \equiv \tau^2$. From (5.141), assuming that the mobility $\mu_{min} = 10^{-6}$ CGS (see [145]), the temperature $T = 300\,\text{K} = 3 \times 10^{-14}$ erg, the electron charge $e = 4.48 \times 10^{-10}$ CGS and the fibril radius $r_{fl} = 10^{-6}$ cm, we get the resultant estimation: $1/\tau_1 = 10^2 1/\text{s}$. If we take the mobility equal to 10^{-2}, the inverse relaxation time appears to be 10^6s^{-1}.

To estimate the time τ_2, we take $n_M = 10^{13}\text{cm}^{-1}$, the temperature $T = 300\text{K} = 3 \times 10^{-14}$ erg, $a = 10^{-8}$ cm, the molecular mass $M = 10^{-25}$ g and the pressure P_2 equal to the atmospheric pressure, i.e. 10^5 erg/cm^3. Then $1/\tau_2 = 10^7\text{s}^{-1}$.

Thus the given numerical calculations prove the correct choice of the relaxation mechanism and frequency range ω. If we increase the frequency ω, we should think about the possible inclusion into the absorption process of new mechanisms of energy dissipation whose time should be less than τ_1 and τ_2. Otherwise, as was stated earlier, it may appear that the energy absorption and together with it the imaginary part of the dielectric permeability ε'' will turn out to be too small, which is never the case physically.

7 Conclusion

The results presented in this monograph are devoted on the whole to a theoretical description of the basic physical parameters of porous dielectrics and pursue only one purpose: to demonstrate the capabilities of the apparatus of theoretical physics for the analysis of such seemingly "disparate" systems as the above-mentioned structures.

The submitted results partially confirm, at least qualitatively, those few experiments at present available which are connected with measurements of the various functional dependences: electric field breakdown, mechanical strength, time delay of ignition and thermal conduction. At the same time so far, unfortunately, there are no experiments on the investigation of sound scattering and electromagnetic radiation in porous media, though the results of such experiments certainly would be important for practical purposes (see Sect. 6.3).

The book suggested to the reader is the first consistent description of the theoretical investigation of the physical parameters and properties of practically very important porous substances and, in particular, of cellulose, its derivatives and paper. We can not but stress that the methods of calculation given in the present monograph may be applied to any other highly disordered structure, for example, composites and compositional materials.

Some more properties and aspects of the porous materials can be found in the following references and we would like to recommend the reader look through these [151–165].

Appendix A: Program for Heat Conductivity Coefficient $\kappa(m, T)$ Calculation

```
PROGRAM HCPS
IMPLICIT DOUBLE PRECISION (A–H, O–Z)
DOUBLE PRECISION CAPA (100), TEMP (100)
DATA IN, IO, PI, IND/5,6,3.1416,1/
DATA XM, DXM, IDXM, XT, DXT, IDXT/0,3,0.,1,60.,0.,1/
DATA V,VMAX,VMIN, VA,VB,RO/1.E−10, 6.E−10, 2.E−10, 1.E−22, 4.E
*  −10, 1.35/
DATA TED,EPS,EPS1,G1,G2,G3,G10/3.E−15, 2.5, 1.E−3, 1.E3,1.E
*  −4, 1.E5,2.E9/
DATA C,CD,CS,H,G4/3.E10,1.E-4,2.E4,1.E-27,1.E9/
DATA DEL,DELTA,P,D,XNN,NY/1.E5,.1.,6.E8,1.E8,1.E−7, 1.E16,100/
DATA YMIN,YMAX,XZ/0.01,1.E2,1./
READ(*,*)XM,XT,G1,G2,G3,GS
WRITE(*,*)XM,DXM,IDXM,XT¡DXT,IDXT,IND
WRITE(*,*)V,VMAX,VMIN,VA,VB,RO
WRITE(*,*) TED,EPS,EPS1,G1,G2,G3,G10
WRITE(*,*) C,CD,CS,H
WRITE(*,*) DEL,DELTA,P,D,XNN,YMIN, YMAX,XZ
XNO=XNN/XM
ZM=VA*RO
XP=60*PI/(G1*G1)
EPSQ=(SQRT(EPS)−0.1)**2
XMM=ZM*CS*CS/TED
VVM=VMAX−VMIN
VV=VVM/G10
XA=5.656*EPS1*EPS1*V*(CD**0.75)
BB1=(1.5/(PI*EPSQ))**0.667
BB2=CS*(XMM**0.333)
BB=442.88*(G3**0.667)*(VA**0.333)*(DEL**5)*G4
BBA=BB1*(TED**4)*BB2*((H*CS)**4)/BBC
AA1=(XP**0.75)*(XMM**1.5)*(VV**0.25)*(CS**1.25)/XA
CC1=TED/(H*C)
CC2=G2*G2*DELTA*TED
CC3=P*D*D
```

```
CCD2=0.1*CC2*(CC1**6)/CC3
CC5=0.667*EPSQ*(CC1**4)/PI
CC7=0.1667*PI*PI*C*DELTA*(CC1**3)
DDB=CD*CS*EPS1*EPS1(XP*(XMM**2)*C)
DD1=(EPS1**2)*(G1**2)*(C**2)*ZM*CD
DD2=(G3**2)*(H**4)*(CS**3)*(DEL**)*(VMAX–VMIN)*(XMM**2)
DD4=(2.9E–4)*DD1/DD2
DO 80 I=1,IDXM
YMM=XM*(1.0–XM)
BBB=(XNN**0.667)*YMM*BBA
AA=AA1*((1.0–XM)**2)
CCD3=CC5/XNO
CC8=CC7*XM*XM
DD5=DD4*YMM
WRITE(*,5) XM
5 FORMAT(2X, XM=',E10.2)
WRITE(*,6)
6 FORMAT(2X,5(2X,'CAPA(J) TEMP(J) ,))
DO 40 J=1,IDXT
XTT=XT/TED
CC9=1.0+CCD2*XTT7+CCD3*XM*XTT
BB=BBB*XTTA
CC=CC8*XTT3/CC9
WRITE(*,6162) DEL,H,C,XT
WRITE(*,6163) TED, XZ, YMIN, YMAX
WRITE(*,6161) NY, IND, IO, DDJ2
CALL INTJ(DEL,H,C,XT,TED,XZ,YMIN,YMAX,NY,IND,IO,DDJ2)
WRITE(*,6162) DEL, H, C, XT
WRITE(*,6163) TED,XZ,YMIN,YMAX
WRITE(*,6161) NY, IND, IO, DDJ2
DD=DD5*XT3*DDJ2
CAPA(J)=AA+BB+CC+DD
TEMP(J)+XT
40 XT=XT+DXT
WRITE(*,8) (CAPA(J)),TEMP(J),J=1,IDXT)
8 FORMAT(2X,10E10.2)
80 XM=XM+DXM
6161 FORMAT (2X,'NY',I10,'IND',I10,'IO',I10,'DDJ2',E17.9)
6162 FORMAT (2X,'DEL',E15.9,'H',E15.9,'C',E15.9,'XT',E15.9)
6163 FORMAT (2X,'TED',E15.9,'XZ',E15.9,'YMIN',E15.9,'YMAX',E15.9)
STOP
END
SUBROUTINE INTJ(DEL,H,C,XT,TED,XZ,YMIN,YMAX,NY,IND,
* IQ,DDJK)
```

```
IMPLICIT DOUBLE PRECISION (A–H,O–Z)
DY=(YMAX-YMIN)/NY
XTT1=XT/H*C
Z1=1/XZ
Z2=Z1/XZ
Z3=1/SQRT(XZ)
Z4=Z3*Z2
S=0.0
I=-1
XY=YMIN
20 I=I+1
XTX=XTT1*XY/DEL
ZZ=XZ+XTX*XTX
ZA=0.125*ZA*DATAAN(XTX*Z3)
ZB=0.125*XTX*Z2/ZZ
ZC=0.834*XTX*Z1(ZZ*ZZ)
ZD=0.333*XTX/(ZZ**3)
F=(ZA+ZB+ZC+ZD)*(XY**4)/(DEXP(XY)-1.0)
XY=XY+DY
IF(I.LT.1) GO TO 30
IF(I.EQ.NY) GO TO 40
S=S+F
GO TO 20
30 S=S+0.5*F
GO TO 20
40 S=S+0.5*F
DDJK=S*DY
IF(IND.EQ.O) GO TO 60
WRITE(*,4) XT,TED,XZ,YMIN,YMAX,DDJK,XY,I,NY
4 FORMAT(3X,7E8.2,215)
60 CONTINUE
RETURN
END
```

Appendix B: Collision Integrals and Relaxation Time Calculation

In order to understand all the reasoning of the present monograph concerning different kinds of relaxation in a substance (whether a porous structure like paper or a much simpler one like, for example, a crystal dielectric with only one pore (see Appendix C)), it is necessary to have an idea of how one can learn to estimate non-equilibrium parameters and to determine the parametric dependence of interest, e.g. that of the thermal conductivity coefficient. Here, as a rule, the most interesting dependence is temperature dependence of κ . . ..

In the present Appendix we shall familiarize the reader with methods of estimation of the relaxation times in the so-called tau-approximation which, however, in spite of its seeming simplicity, allows us semi-phenomenologically to find the correct temperature behavior of most of the macroscopic non-equilibrium parameters. At least, this approximation (as we shall rigorously prove mathematically) appears to be quite sufficient to describe the thermal conductivity coefficient of a crystal dielectric *over the whole real* range of temperature changes. And here it is necessary to pay attention to the following. Theoretically, the thermal conductivity coefficient, as a function of T, has been estimated in a great number of original papers, but they all have one and the same drawback: they analytically describe either the range of very low (helium) temperatures or the range of very high temperatures exceeding that of Debye, notwithstanding the fact that the most interesting range (both from the theoretical and experimental point of view) is undoubtedly the range in the maximum region of the coefficient of thermal conductivity. It is quite understandable: the fact is that it is impossible to find out analytically how $\kappa(T)$ behaves near a maximum, because for this purpose it is necessary to apply only numerical methods of calculation. That is why the present Appendix has not only an instructive character, but it also pursues the aim to teach the reader methods of work with non-equilibrium parameters. Besides, touching upon the description of the dielectric thermal conductivity coefficient in the real temperature range (from absolute zero up to the Debye temperature), we shall logically complete the issue raised so far and show how problems of this type can be solved.

Here it is important to underline initially that the ultimate mechanism for establishing of thermodynamic equilibrium among all the subsystems participating in interactions (longitudinal, transverse and optical phonons) is their connection with the thermostat. We shall see below that this very mechanism is the last stage of relaxation and it determines the true temperature dependence of the thermal conductivity coefficient.

All the calculations will be performed in the so-called "tau-approximation" by means of the quasi-classical kinetic Boltzmann equation, whose range of applicability was discussed in Sect. 4.7.

The first step is to write the invariant Hamiltonian (relative to the transformation group of the given crystal), and for strictional interaction in case of arbitrary (so far) symmetry we have

$$H_{\text{int}} = \frac{\Theta_D}{V} \gamma_{iklmnp} \int_V u_{ik} u_{lm} u_{np} \mathrm{d}^3 x, \tag{B.1}$$

where V is the dielectric volume and deformation tensor is

$$u_{ik} = \frac{1}{2} \left(\frac{\partial u_i}{\partial x_k} + \frac{\partial u_k}{\partial x_i} \right), \tag{B.2}$$

where u_i are the components of the shift vector $\boldsymbol{u}(x, y, z)$, γ_{iklmnp} are the dimensionless components of the striction tensor in six dimensions, whose symmetry properties are given by equation (B.1), and the relation with the experimentally determined denominate quantities C_{iklmnp} (see for example [138]) is given by $\gamma_{iklmnp} = C_{iklmnp}(a^3/6\Theta_D)$. The indices that are repeated in (B.1) mean summation.

In the general case the symmetry tensor γ_{iklmnp} may have up to 56 independent components. Even for crystal cubic symmetry $m3m$ there are 6 components, and in isotope case there are 3. That is why we shall determine an effective constant $\bar{\gamma}$ by presenting the tensor γ_{iklmnp} in isotropic form with one independent component (for cubic symmetry crystals it will be the most realistic case). The value $\bar{\gamma}$ may be determined from the compliance of the theoretical value of the thermal conductivity coefficient κ with the experimental data. An isotropic tensor in six dimensions may be presented obviously as a linear combination of $6!/(2^3 \cdot 3!) = 15$ isomers. Taking into account the symmetry of the tensor γ_{iklmnp} and the above-mentioned allowances we obtain the following equation:

$$\begin{aligned}
\gamma_{iklmnp} = g \cdot [& \delta_{np}(\delta_{ik}\delta_{lm} + \delta_{il}\delta_{km} + \delta_{kl}\delta_{im}) \\
& + \delta_{il}(\delta_{kn}\delta_{mp} + \delta_{kp}\delta_{nm}) + \delta_{km}(\delta_{in}\delta_{pl} + \delta_{ip}\delta_{nl}) \\
& + \delta_{ik}(\delta_{ln}\delta_{mp} + \delta_{lp}\delta_{nm}) + \delta_{lm}(\delta_{in}\delta_{kp} + \delta_{ip}\delta_{kn}) \\
& + \delta_{kl}(\delta_{in}\delta_{mp} + \delta_{ip}\delta_{nm}) + \delta_{im}(\delta_{ln}\delta_{kp} + \delta_{lp}\delta_{kn})].
\end{aligned} \tag{B.3}$$

Simplifying it over indices $l-k$, $n-m$, $p-i$ we find $\gamma_{ikknni} = 27\bar{\gamma} = 105\,\mathrm{g}$, where $\bar{\gamma}$ is obtained as the arithmetic mean of the 27 terms of the equation $\gamma_{ikknni} \equiv \sum_{i,k,n=1}^{3} \gamma_{ikknni}$. Consequently, $g = (9/35) \cdot \bar{\gamma}$ and the Hamiltonian (B.1) will be reduced to the resultant form:

$$H_{\text{int}} = \frac{9}{35}\bar{\gamma}\frac{\Theta_D}{V}\int\limits_V \left(u_{ii}^3 + 6u_{ik}^2 u_{nn} + 8u_{ik}u_{kn}u_{ni}\right) \cdot \mathrm{d}^3 x \,. \tag{B.4}$$

Writing now the shift vector $\boldsymbol{u}$ in the secondary-quantized representation, that is

$$\boldsymbol{u} = \sum_{k,j}\left(\frac{\hbar}{2\rho V\omega_{jk}}\right)^{1/2}\left(b_{kj}^+ + b_{-kj}\right)\cdot \boldsymbol{e}_j \cdot \exp\left(i\boldsymbol{kx}\right), \tag{B.5}$$

where ρ is dielectric density, and $\boldsymbol{e}_j$ is polarization vector with components $j = 1,2,3$; we may get the deformation tensor components. Actually,

$$u_{nm} = \frac{\mathrm{i}}{2}\sum_{k,j}\left(\frac{\hbar}{2\rho V\omega_{jk}}\right)^{1/2}(k_n e_{jm} + k_m e_{jn})\cdot\left(b_{kj}^+ + b_{-kj}\right)\cdot\exp\left(\mathrm{i}\boldsymbol{kx}\right)\,. \tag{B.6}$$

Substituting them into (B.4) we obtain

$$H_{\text{int}} = \sum_{\{\boldsymbol{k}\},\{j\}} \Psi_{\{j\}}(\{\boldsymbol{k}\})\cdot(b_{k_1 j}^+ + b_{-k_1 j})\cdot(b_{k_2 j'}^+ + b_{-k_2 j'})$$
$$\times(b_{k_3 j''}^+ + b_{-k_3 j''})\Delta(\boldsymbol{k}_1 + \boldsymbol{k}_2 + \boldsymbol{k}_3)\,, \tag{B.7}$$

where $\{\boldsymbol{k}\} = \boldsymbol{k}_1, \boldsymbol{k}_2, \boldsymbol{k}_3$; $\{j\} = j, j', j''$, and the scattering amplitude is

$$\Psi_{\{j\}}(\boldsymbol{k}_1, \boldsymbol{k}_2, \boldsymbol{k}_3) = -\mathrm{i}\frac{9\bar{\gamma}\Theta_D}{35}\left(\frac{\hbar}{2\rho V}\right)^{3/2}$$
$$\times\frac{(k_1 e_j)\cdot\left[(k_2 e_{j'}')(k_3 e_{j''}'') + 6(k_2 e_j'')(k_3 e_j')\right]}{(\omega_{k_1 j}\omega_{k_2 j'}\omega_{k_3 j''})^{1/2}}\,, \tag{B.8}$$

$b_{kj}^+ (b_{kj})$ is the creation (annihilation) operator of a phonon with wave vector $\boldsymbol{k}$ and polarization $\boldsymbol{e}_j$.

It may be shown that the laws of energy and momentum conservation admit only the following four dissipative processes:

$(1) b_{lk}^+ b_{tk_1}^+ b_{lk_2}$; $(2) b_{lk}^+ b_{tk_1} b_{tk_2}$; $(3) b_{lk}^+ b_{tk'}$; $(4) b_{lk}^+ b_{ok_1} b_{ok_2}^+$,

where the indices "t" and "l" refer to transverse and longitudinal phonons, respectively, and the index "o" refers to optical phonons. For process (1) the allowed values of the virtual wave vector are determined by the inequality:

$$0 \le k_1 \le \frac{2k\lambda}{(\lambda - 1)}\,, \tag{B.9}$$

where $\lambda = c_l/c_t$.

For process (2):

$$0.5k(\lambda - 1) \le k_1 \le 0.5k(\lambda + 1)\,. \tag{B.10}$$

Finally, for process (4):

$$\frac{\pi}{a} \geq k_1 \geq \frac{1}{2}\left(\frac{C_l}{\beta} + k\right),$$ (B.11)

where β in introduced from the definition of the optical phonons spectrum $\omega_{0k} = \omega_0 - \beta k^2$, which is valid in the long-wave approximation, on condition that the product $a \cdot k$ is small.

As an illustrative example let us show how to find the inequality (B.9).

For the interaction $b_{lk}^+ b_{tk_1}^+ b_{lk_2}$ we have two conservation laws. The conservation law for momentum is

$$\boldsymbol{k} + \boldsymbol{k}_1 - \boldsymbol{k}_2 = 0$$

and the conservation law for energy is

$$\omega_{lk} + \omega_{tk_1} - \omega_{lk_2} = 0.$$

As $\omega_{lk} = c_l k$ and $\omega_{tk} = c_t k$, we have

$$c_l k + c_t k_1 - c_l k_2 = 0.$$

Introducing the dimensionless parameter $\xi = c_t/c_l$, we obtain

$$k + \xi k_1 = k_2.$$

From the conservation momentum law we can express $\boldsymbol{k}_2$ as $\boldsymbol{k}_2 = \boldsymbol{k} + \boldsymbol{k}_1$. Hence we have

$$k + \xi k_1 = |\boldsymbol{k} + \boldsymbol{k}_1|.$$

If we raise to the second power this expression and introduce the angle θ between vectors $\boldsymbol{k}$ and $\boldsymbol{k}_1$ we obtain

$$k^2 + 2kk_1\xi + k^2\xi^2 = k^2 + 2kk_1\cos\theta + k_1^2.$$

From this we obtain

$$\cos\theta = \frac{(\xi^2 - 1)k_1 + 2k\xi}{2k}.$$

Because $|\cos\theta| \leq 1$ we receive the following double inequality:

$$-1 \leq \frac{(\xi^2 - 1)k_1 + 2k\xi}{2k} \leq 1.$$

From this we obtain

$$\begin{cases} (\xi^2 - 1)k_1 + 2k(\xi - 1) \leq 0 \quad \text{and} \\ (\xi^2 - 1)k_1 + 2k(\xi + 1) \geq 0. \end{cases}$$

As $\xi < 1$, we can write

$$\begin{cases} (1 - \xi^2)k_1 + 2k(1 - \xi) \geq 0 \quad \text{and} & (A) \\ (1 - \xi^2)k_1 - 2k(1 + \xi) \leq 0. & (B) \end{cases}$$

After cancellation of the inequality (A) on $(1 - \xi)$ we see that in the final result it is automatically then implemented. And (B) is given by

$$k_1 \leq \frac{2k}{(1 - \xi)}.$$

Substituting here the inverse value $\lambda = 1/\xi$ we obtain the inequality (B.9), which was to be proved.

Accordingly, in the Hamiltonian (B.7) only three terms should be left. They are

$$H_{\text{int}} = \sum_{\{k\}} \sum_{\{j\}} [\Psi_1^{\{j\}}(\{\boldsymbol{k}\}) \cdot b_{lk_1}^+ b_{lk_2} b_{tk_3}^+ \cdot \Delta(\boldsymbol{k}_1 - \boldsymbol{k}_2 + \boldsymbol{k}_3) + \text{C.C.}]$$

$$+ \sum_{\{k\}} \sum_{\{j\}} [\Psi_2^{\{j\}}(\{\boldsymbol{k}\}) \cdot b_{lk_1}^+ b_{tk_2} b_{tk_3} \cdot \Delta(\boldsymbol{k}_1 - \boldsymbol{k}_2 - \boldsymbol{k}_3) + \text{C.C.}]$$

$$+ \sum_{\{k\}} \sum_{\{j\}} [\Psi_3^{\{j\}}(\{\boldsymbol{k}\}) \cdot b_{lk_1}^+ b_{ok_2} b_{ok_3}^+ \cdot \Delta(\boldsymbol{k}_1 - \boldsymbol{k}_2 + \boldsymbol{k}_3) + \text{C.C.}],$$

$$(B.12)$$

where C.C. is the complex conjugate and the scattering amplitude is

$$\Psi_1^{(j)}(\boldsymbol{k}_1, \boldsymbol{k}_2, \boldsymbol{k}_3) = -\frac{9i\bar{\gamma}\Theta_D}{35} \left(\frac{\hbar}{2\rho V}\right)^{3/2} \frac{e_z' k_{1z}}{\sqrt{\omega_{lk_1}\omega_{lk_2}\omega_{tk_3}}}$$

$$\times [(\boldsymbol{e}_\perp'' \boldsymbol{k}_{3\perp}) \cdot k_{2z} e_z' + 6k_{3z} e_z''(\boldsymbol{k}_{2\perp} \boldsymbol{e}_\perp')] ; \qquad (B.13)$$

$$\Psi_2^{(j)}(\boldsymbol{k}_1, \boldsymbol{k}_2, \boldsymbol{k}_3) = -\frac{9i\bar{\gamma}\Theta_D}{35} \left(\frac{\hbar}{2\rho V}\right)^{3/2} \frac{k_{1z} e_z}{\sqrt{\omega_{lk_1}\omega_{tk_2}\omega_{tk_3}}}$$

$$\times \{(\boldsymbol{k}_{2\perp} \boldsymbol{e}_\perp') \cdot (\boldsymbol{k}_{3\perp} \boldsymbol{e}_\perp'') + 3 \cdot [(\boldsymbol{k}_2 \boldsymbol{e}_\perp'') \cdot (\boldsymbol{k}_{3\perp} \boldsymbol{e}_\perp')$$

$$+ (\boldsymbol{k}_{2\perp} \boldsymbol{e}_\perp') \cdot (\boldsymbol{k}_{3\perp} \boldsymbol{e}_\perp'')]\} ; \qquad (B.14)$$

$$\Psi_3^{(j)}(\boldsymbol{k}_1, \boldsymbol{k}_2, \boldsymbol{k}_3) = -\frac{9i\bar{\gamma}\Theta_D}{35\omega_0\sqrt{C_l}}(\frac{\hbar}{2\rho V})^{3/2}$$

$$\times \frac{(k_{1z} e_z) \cdot [(\boldsymbol{k}_2 \boldsymbol{e}') \cdot (\boldsymbol{k}_3 \boldsymbol{e}'') + 6(\boldsymbol{k}_2 \boldsymbol{e}'') \cdot (\boldsymbol{k}_3 \boldsymbol{e}')]}{\sqrt{k_1}}. \qquad (B.15)$$

Note, that the scattering amplitude $\Psi_2^{(j)}$ is symmetrical with respect to substitution of $\boldsymbol{k}_2$, $\boldsymbol{e}'$ with $\boldsymbol{k}_3$, $\boldsymbol{e}''$, i.e.,

$$\Psi_2^{(j)}(\boldsymbol{k}_1, \boldsymbol{k}_2, \boldsymbol{k}_3, e, e', e'') = \Psi_2^{(j)}(\boldsymbol{k}_1, \boldsymbol{k}_3, \boldsymbol{k}_2, e, e'', e').$$

Let include the impurity scattering of phonons. For its estimation the elastic interaction of phonons with impurity atoms should be written down. In phenomenological form it may be presented in the following way:

$$H_{\text{int}} = \sum_{\boldsymbol{k}_1, \boldsymbol{k}_2} \Psi_{\text{imp}}^{(j)}(\boldsymbol{k}_1, \boldsymbol{k}_2) \cdot b_{lk_1 j}^+ b_{tk_2 j'} , \tag{B.16}$$

where the elastic scattering amplitude is

$$\Psi_{\text{imp}}^{(j)}(\boldsymbol{k}_1, \boldsymbol{k}_2) = \frac{\sqrt{c_i} \cdot \bar{\gamma}_{\text{imp}} \Theta_D \hbar \cdot (e_{1z} k_{1z}) \cdot (e_{2\perp} \boldsymbol{k}_{2\perp})}{2\rho V \sqrt{C_l C_t} \sqrt{k_1 k_2}} , \tag{B.17}$$

where $\bar{\gamma}_{\text{imp}}$ is a dimensionless constant of the interaction between the impurities and the lattice, and c_i is a dimensionless impurity concentration defined as the ratio of the number of impurity atoms to the total number of dielectric atoms.

By means of Hamiltonians (B.12) and (B.16) we may write the corresponding collision integral. The recipe for writing collision integrals in the quasi-classical approximation, i.e. $kl \gg 1$, is known, and using the algorithm of the corresponding formal procedure described, say, in monographs [126, 139] (see also the review [54]), we arrive at the following equation:

$$L\{N_{lk}\} = L_1\{N_{lk}\} + L_2\{N_{lk}\} + L_3\{N_{lk}\} + L_{\text{imp}}\{N_{lk}\} , \tag{B.18}$$

where the collision integrals are

$$L_1\{N_{lk}\} = \frac{2\pi}{\hbar^2} \sum_{\boldsymbol{k}_{1,2}} \sum_{e', e''} \left| \Psi_1^{(j)}(\boldsymbol{k}, \boldsymbol{k}_1, \boldsymbol{k}_2) \right|^2$$
$$\times \Delta(\boldsymbol{k} - \boldsymbol{k}_1 + \boldsymbol{k}_2) \cdot \delta(\omega_{lk} - \omega_{lk_1} + \omega_{tk_2})$$
$$\times \left[(1 + N_{lk}) N_{lk_1} (1 + N_{tk_2}) - N_{lk} N_{tk_2} (1 + N_{lk_1}) \right] , \tag{B.19}$$

$$L_2\{N_{lk}\} = \frac{2\pi}{\hbar^2} \sum_{\boldsymbol{k}_{1,2}} \sum_{e', e''} \left| \Psi_2^{(j)}(\boldsymbol{k}, \boldsymbol{k}_1, \boldsymbol{k}_2) \right|^2$$
$$\times \Delta(\boldsymbol{k} - \boldsymbol{k}_1 - \boldsymbol{k}_2) \cdot \delta(\omega_{lk} - \omega_{tk_1} - \omega_{tk_2})$$
$$\times \left[(1 + N_{lk}) N_{tk_1} N_{tk_2} - N_{lk} (1 + N_{tk_1}) \cdot (1 + N_{tk_2}) \right] , \tag{B.20}$$

$$L_3\{N_{lk}\} = \frac{2\pi}{\hbar^2} \sum_{\boldsymbol{k}_{1,2}} \sum_{e', e''} \left| \Psi_3^{(j)}(\boldsymbol{k}, \boldsymbol{k}_1, \boldsymbol{k}_2) \right|^2$$
$$\times \Delta(\boldsymbol{k} - \boldsymbol{k}_1 + \boldsymbol{k}_2) \cdot \delta(\omega_{lk} - \omega_{ok_1} + \omega_{ok_2})$$
$$\times \left[(1 + N_{lk}) N_{ok_1} (1 + N_{ok_2}) - N_{lk} (1 + N_{ok_1}) \cdot N_{ok_2} \right] , \tag{B.21}$$

$$L_{\text{imp}}\{N_{lk}\} = \frac{2\pi}{\hbar^2} \sum_{\boldsymbol{k}', e'} \left| \Psi_{\text{imp}}^{(j)}(\boldsymbol{k}, \boldsymbol{k}') \right|^2 (N_{tk'} - N_{lk}) \cdot \delta(\omega_{tk'} - \omega_{lk'}) . \tag{B.22}$$

Now we show the algorithm of receipt collision integrals (B.19)–(B.22).

Let us assume a Hamiltonian of the type $H_{\text{int}} = \sum (\psi b_1 b_2^+ b_3^+ + \psi^* b_1^+ b_2 b_3)$ and let us introduce the following designations:

$$\begin{cases} h = \psi b_1 b_2^+ b_3^+ & \text{and} \\ h^+ = \psi^* b_1^+ b_2 b_3 \,. \end{cases}$$

According to Fermi's "golden rule" the probability of transition per unit time is

$$W = \left(\frac{2\pi}{\hbar}\right) \sum_m |\langle n|h_{\text{int}}|m\rangle|^2 \delta(E_n - E_m)\,,$$

where the indexes "n" and "m" describe the quantum states of the system.

In this case the role of the quantum states play momentum $\boldsymbol{k}$ and h_{int} is operators h and h^+.

For the non-equilibrium process we must not write but W and difference between fluxes coming to W^+ and thinging from W^-, i.e. $W = W^+ - W^-$, where

$$\begin{cases} W^+ = \left(\dfrac{2\pi}{\hbar}\right) \sum_{\nu'} |\langle \nu|h|\nu'\rangle|^2 \delta(\sum \varepsilon)\,, & \text{and} \\ W^- = \left(\dfrac{2\pi}{\hbar}\right) \sum_{\nu'} |\langle \nu|h^+|\nu'\rangle|^2 \delta(\sum \varepsilon_\nu)\,. \end{cases}$$

The matrix element of h to the second power is

$$|\langle |h|\rangle|^2 = |\psi|^2 \langle b_1 b_2^+ b_3^+ b_1^+ b_2 b_3 \rangle = |\psi|^2 \langle b_1 b_1^+ \rangle \langle b_2^+ b_2 \rangle \langle b_3^+ b_3 \rangle \,.$$

As

$$\begin{cases} \langle b_1 b_1^+ \rangle = 1 + N_{k_1}\,, \\ \langle b_2^+ b_2 \rangle = N_{k_2}\,, \\ \langle b_3^+ b_3 \rangle = N_{k_3}\,, \end{cases}$$

we have

$$|\langle |h|\rangle|^2 = |\psi|^2 (1 + N_{k_1}) N_{k_2} N_{k_3}$$

and analogously

$$|\langle |h^+|\rangle|^2 = |\psi|^2 N_{k_1} (1 + N_{k_2})(1 + N_{k_3})\,.$$

Hence, their difference is the collision integral (B.20), which was to be proved. The other collision integrals are obtained analogously.

As to the relaxation time, we shall estimate it in the τ-approximation, which is sufficient for our purposes, by the formula:

$$\frac{1}{\tau_{lk}} = - \left.\frac{\delta L}{\delta N_{lk}}\right|_{\substack{N_{lk}=\bar{N}_{lk} \\ N_{tk}=\bar{N}_{tk}}}\,, \tag{B.23}$$

where the equilibrium distribution functions $\bar{N}_{l,tk}$ are the conventional Bose distribution functions, i.e. $\bar{N}_{l,tk} = (\exp\{\hbar\omega_{l,t,k}/T\} - 1)$.

228   Appendix B

The total relaxation time is the sum of the times

$$\frac{1}{\tau_{lk}} = \frac{1}{\tau_{1lk}} + \frac{1}{\tau_{2lk}} + \frac{1}{\tau_{3lk}} + \frac{1}{\tau_{impk}} + \frac{C_l}{R}, \tag{B.24}$$

where each of the times is calculated individually for the corresponding interaction process.

The variation derivative of the collision integral in the formula (B.23), is found rather simply by taking account of the explicit expressions for the collision integrals (B.19)–(B.23).

In order to demonstrate the method of finding the inverse relaxation time in the τ-approximation let us choose the time $1/\tau_{1lk}$ and show from this example what the algorithm of the calculation should be. For this purpose let us consider the collision integral L_1 from the equation (B.19).

According to the definition (B.23), we find the following expression:

$$\frac{1}{\tau_{1ltk}} = \frac{2\pi}{\hbar^2} \sum_{k_{1,2}} \sum_{e',e''} \left| \Psi_1^{(j)}(\boldsymbol{k}, \boldsymbol{k}_1, \boldsymbol{k}_2) \right|^2$$

$$\times \Delta(\boldsymbol{k} - \boldsymbol{k}_1 + \boldsymbol{k}_2) \cdot \delta(\omega_{lk} - \omega_{lk_1} + \omega_{tk_2})[\bar{N}_t(k_2) - \bar{N}_l(k_1)].$$

This reverse time is obviously the reverse time $1/\tau_{1lk}$.

Let us demonstrate in more detail how this equality is found. So, from the rule (B.23) and expressions (B.19) we have

$$\frac{\delta L}{\delta N_{lk}} = \frac{2\pi}{\hbar^2} \sum_{k_{1,2}} \sum_{e''} \left| \Psi_1^{(j)}(\boldsymbol{k}, \boldsymbol{k}_1, \boldsymbol{k}_2) \right|^2$$

$$\times \Delta(\boldsymbol{k} - \boldsymbol{k}_1 + \boldsymbol{k}_2) \cdot \delta(\omega_{lk} - \omega_{lk_1} + \omega_{tk_2})$$

$$\times \left(\frac{\delta}{\delta N_{lk}} \right) [(1 + N_{lk})N_{lk_1}(1 + N_{tk_2}) - N_{lk}N_{tk_2}(1 + N_{lk_1})].$$

Because

$$\left(\frac{\delta}{\delta N_{lk}} \right) [(1 + N_{lk})N_{lk_1}(1 + N_{tk_2}) - N_{lk}N_{tk_2}(1 + N_{lk_1})]$$

$$= N_{lk_1}(1 + N_{tk_2}) - (1 + N_{lk_1})N_{tk_2} = N_{lk_1} - N_{tk_2},$$

we immediately found the above-mentioned expression.

So far, the analysis of the energy and momentum conservation laws in the obtained equation is contained in the joint solution of two equations:

$$\begin{cases} \boldsymbol{k} - \boldsymbol{k}_1 + \boldsymbol{k}_2 = 0, \\ \omega_l(k) - \omega_l(k_1) + \omega_t(k_2) = 0, \end{cases}$$

where the frequencies are $\omega_l(k) = c_l k$, $\omega_t(k) = c_t k$.

The simplest analysis (see above) of this system of equations allows us to determine the regions of wave vector changes, which will be necessary for integration in virtual k-space and which as a result of the simplest algebraic calculations turn out to be the following (see (B.9)):

$$k \leq k_1 \leq \frac{k(\lambda + 1)}{(\lambda - 1)},$$

$$0 \leq k \leq \frac{\pi}{a}.$$

We see that the region of k-integration is limited by the first Brillouin zone. The transition from summation to integration in the formula for the inverse relaxation time is performed by the introduction of the number of states and the formal substitution $\sum(\ldots) = V \int(\ldots)d^3k/(2\pi)^3$.

Finally, after summation over the virtual wave vectors, which is greatly simplified due to the presence of a delta function expressing the law of energy conservation, and using the rule of averaging over directions of phonon polarization we have $\langle e_i e_j \rangle = \delta_{ij}/3$, where δ_{ij} is the Kronecker delta, we find that

$$\frac{1}{\tau_{1lk}} = \tilde{\gamma}_1 \frac{\Theta_D^2 \hbar a^3}{M^3 C_l^2 C_t^2} \cdot \int\limits_k^{\alpha k} dk_1 \left[\bar{N}_{l(k_1-k)} - \bar{N}_{lk_1} \right]$$

$$\times \left\{ k_1^2 A^2 [(k_1 - k)^2 \lambda^2 - (k_1 A - k)^2] + 6 k_1^4 (1 - A^2)^2 \right.$$
$$\left. + 36 k_1^2 (1 - A^2)(k_1 A - k)^2 \right\}. \tag{B.25}$$

Here,

$$\tilde{\gamma}_1 = \bar{\gamma}^2 \frac{1}{39^2 \cdot 108 \cdot \pi} \approx 1.94 \times 10^{-6} \bar{\gamma}^2,$$

$$A = \frac{k^2 + k_1^2 - \lambda^2 (k - k_1)^2}{2kk_1} \quad \text{and obviously} A \leq 1,$$

$$\lambda = \frac{C_l}{C_t} \geq 1, \quad \alpha = \frac{(\lambda + 1)}{(\lambda - 1)}.$$

The rest times may be found in absolute compliance with the given rule and, as a result, the rest relaxation times defined by the collision integrals L_2, L_3 and L_{imp} will be

$$\frac{1}{\tau_{2lk}} = \tilde{\gamma}_2 \frac{\Theta_D^2 \hbar a^3}{M^3 C_l C_t^3} \cdot \int\limits_{z_1 k}^{z_2 k} k_1^4 (1 - B^2) dk_1 \left[1 + \bar{N}_{tk_1} - \bar{N}(\omega_{lk} - \omega_{tk_1}) \right], \tag{B.26}$$

where

$$\tilde{\gamma}_2 = \bar{\gamma}^2 \frac{49}{39^2 \cdot 108 \cdot \pi} \approx 9.5 \times 10^{-5} \gamma^2,$$

$$B = \frac{k(1 - \lambda^2) + 2k_1 \lambda}{2k_1},$$

$$z_1 = \frac{(\lambda - 1)}{2},$$

$$z_2 = \frac{(\lambda + 1)}{2}.$$

$$\frac{1}{\tau_{3k}} = \tilde{\gamma}_3 \frac{\Theta_D^2 \hbar a^3}{M^3 C_l^\beta \omega_o^2} \cdot \int\limits_{z_3}^{\infty} k_1 \left[\bar{N}_{ok_1} - \bar{N}_o(\omega_{lk} - \omega_{ok_1}) \right]$$

$$\times \left\{ k_1^4 - \frac{C_l}{\beta} k k_1^2 + \frac{6}{49} k^2 \left(\frac{C_l}{\beta} + k \right)^2 - \frac{20}{49} k^2 k_1^2 \right\} dk_1 , \qquad (B.27)$$

where

$$z_3 = \frac{1}{2} \left(k + \frac{C_l}{\beta} \right) ,$$

and the parameter β is defined by the equality

$$\omega_{ok} = \omega_o - \beta k^2 . \qquad (B.28)$$

Its dimensions are $[\text{cm}^2/\text{s}]$; according to the simple theory of crystal lattice atom oscillations [140], for a cubic crystal we may take $\omega_o \cong 2(C_S/a)$ and $\beta \approx a C_S/4$.

Then, constant

$$\tilde{\gamma}_3 = \bar{\gamma}^2 \frac{49}{39^2 \cdot 32 \cdot 27} \approx 3.7 \times 10^{-5} \bar{\gamma}^2 ,$$

and the impurity relaxation time, therefore, will be

$$\frac{1}{\tau_{\text{impk}}} = \frac{\pi c_i \bar{\gamma}_{\text{imp}}^2 \Theta_D^2 a^3 \lambda^4 k^4}{18 \cdot M^2 C_t^2 C_l} . \qquad (B.29)$$

The time L/c_l takes into account the so-called Knudsen scattering of longitudinal phonons by the sample boundaries (L is the linear dimension of a sample). This time begins to play an important role only in the range of low temperatures when the process of phonon thermalization is caused by its collision with the dielectric boundary.

This relaxation time is mainly responsible for the widening of the thermal conductivity coefficient curve $\kappa(T)$.

By means of the formulae (B.24)–(B.27) and (B.29) the thermal conductivity coefficient in the gas-kinetic approximation may be written in a rather compact form that is also very convenient for analysis. Indeed, for reference we shall give the complete expression for $\kappa(T)$, in which all the main relaxation mechanisms are accounted for:

$$\kappa(T, \lambda, R) = \frac{R}{a} \left(\frac{T}{\Theta_D} \right)^3 \frac{\pi}{2 \cdot 3^{1/3}} \cdot \frac{C_l}{a^2 (1 + 2\lambda^3)^{2/3}}$$

$$\times \int\limits_0^{\Theta_D^*/T} \frac{y^4 \bar{N}_y (1 + \bar{N}_y)}{1 + (\delta/L) F(y)} dy , \qquad (B.30)$$

where

$$\Theta_D^* = \Theta_D \left(\frac{3}{1 + 2\lambda^3}\right)^{1/3},$$

and the function

$$F(y) = F_1(y) + F_2(y) + F_3(y) + F_{\text{imp}}(y). \tag{B.31}$$

Here every component is presented in the form:

$$F_1(y) = \tilde{\gamma}_1^* \frac{R}{a} \left(\frac{\hbar}{\rho a^4 C_l}\right)^3 \frac{\lambda^2(\lambda^2 - 1)}{(1 + 2\lambda^3)^2} y^5 \left(\frac{T}{\Theta_D}\right)^5 \left(\frac{C_l}{C_S}\right)$$

$$\times \int_1^\alpha (x - 1)^2 \varphi_1(x) \left[\bar{N}(y(x-1)) - \bar{N}(xy)\right] \mathrm{d}x, \tag{B.32}$$

where,

$$\tilde{\gamma}_1^* = \frac{3 \cdot \pi^6 \bar{\gamma}^2}{39^2 \cdot 16} \approx 0.12 \bar{\gamma}^2.$$

The functions φ_1 and F_2 are given by

$$\varphi_1(x) = [x^2 + 1 - (x-1)^2 \lambda^2][(x+1)^2 - \lambda^2(x-1)^2]$$
$$+ 6(x-1)^2(\lambda^2 - 1)[x^2 + 1 - \lambda^2(x-1)^2]^2$$
$$+ 36(x-1)^2[(x+1)^2 - \lambda^2(x-1)^2]^3,$$

$$F_2(y) = \tilde{\gamma}_2^* \frac{R}{a} \left(\frac{\hbar}{\rho a^4 C_l}\right)^3 \frac{\lambda^3(\lambda^2 - 1)^2}{(1 + 2\lambda^3)^2} y^5 \left(\frac{T}{\Theta_D}\right)^5 \left(\frac{C_l}{C_S}\right)$$

$$\times \int_{z_1}^{z_2} \varphi_2(x) \left[1 + \bar{N}\left(\frac{xy}{\lambda}\right) - \bar{N}\left(y\left(1 - \frac{x}{\lambda}\right)\right)\right] \mathrm{d}x, \tag{B.33}$$

where

$$\tilde{\gamma}_2^* = \frac{49}{32} \tilde{\gamma}_1^* \approx 0.18 \bar{\gamma}^2,$$

$$\varphi_2(x) = \left(x - \frac{\lambda+1}{2}\right)^2 \left(x - \frac{\lambda-1}{2}\right)^2.$$

As $F_3(y) = (L/c_S) \cdot (1/\tau_{3k})$, then, taking into account (B.27) and the optical phonon dispersion (B.28), we find that at $T \ll \Theta_D$

$$F_3(y) \cong \tilde{\gamma}_3^* \frac{R}{a} \left(\frac{\hbar}{\rho a^4 C_l}\right)^3 (1 + 2\lambda^3)^{2/3} \left(\frac{T}{\Theta_D}\right)^3 (1 - \exp(-y))$$

$$\times \exp\left(-\frac{2\Theta_D}{\pi T}\right) \left(\frac{C_l}{C_S}\right) \int_{X_H(y)}^\infty \varphi_3(x) \sqrt{x} \exp(-x) \mathrm{d}x, \tag{B.34}$$

where the lower integration limit is

$$X_H(y) = \frac{f}{4}\left(y + \frac{1}{f}\right)^2,$$

$$\varphi_3(x, y) = x^2 - xy - \frac{20}{49}fxy^2 + \frac{5}{49}(fy)^2\left(y + \frac{1}{f}\right)^2,$$

$$f = \frac{3\pi}{4(1 + 2\lambda^2)}\frac{T}{\Theta_D},$$

$$\tilde{\gamma}_3^* = \frac{49 \cdot \pi^4 \bar{\gamma}^2}{81 \cdot 39^2} \approx 0.04\gamma^2.$$

$$F_{\text{imp}}(y) = c_i B_i y^4 \left(\frac{T}{\Theta_D}\right)^4 \left(\frac{C_l}{C_S}\right), \tag{B.35}$$

where

$$B_i = \frac{\pi}{18}\frac{R}{a}\lambda^5 \left(\frac{\bar{\gamma}_{\text{imp}}\Theta_D}{M C_l^2}\right)^2. \tag{B.36}$$

Here it should be noted that the formula (B.30) will "work" only if the impurity concentration is relatively low.

If these impurities are available, the laws of energy and momentum conservation allow the elastic transformation process of a longitudinal phonon into a transverse thermostat phonon, and thus, the time of this process τ_{imp} may compete (at relatively high impurity concentration) with the time τ_{llt}. It means that in calculating κ both relaxation mechanisms should be included in parallel. The establishment of a genuine equilibrium temperature T_0 in the longitudinal and transverse phonon systems is achieved due to their "contact" with the phonons of the thermostat that maintains the temperature T; the given process proceeds in parallel with the processes H_{lltint} and H_{lttint}.

For the optical phonons, the last term in the Hamiltonian (B.12) is responsible for the process of "adjustment" of the temperature of the optical phonons to the temperature of the longitudinal phonons T_t. The corresponding time is $\tau_{\text{opt}-l}$.

All the above (if we abstract from the thermostat!!) may be united by the following chain of inequalities:

$$\tau_{t-t} \ll \tau_{l-t}, \quad \tau_{l-\text{opt}}, \quad \tau_{l-\text{imp}} \ll \tau_{l-l}. \tag{B.37}$$

The relaxation theory is described in detail in Sect. 4.7.

The numerical analysis of the general expression for the thermal conductivity coefficient is shown in Figs. B.1 and B.2.

Numerical integration of formulae (B.25)–(B.27) and (B.29) allows us to obtain the graphic relationship between the relaxation times illustrated in Fig. B.3.

Completing this Appendix and concluding it logically, we would like to underline once more the main positions as follows.

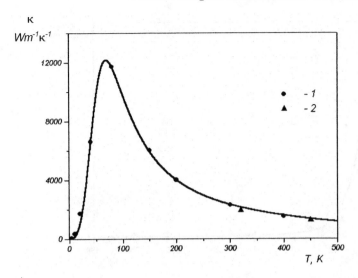

a)

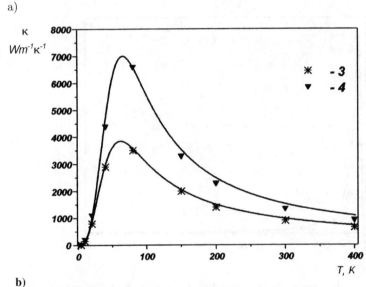

b)

Fig. B.1. (**a**) Dependence of diamond thermal conductivity coefficient on the temperature. The triangles and points correspond to experiments of different authors included in the Encyclopedia of Physics [138] solid line is the theoretical calculation according to the formula (B.30) given in Supplement 1. (**b**) The correspondence between the theoretical calculations according to the same formula (B.30) with the application of the numerical integration technique (for which the author is very grateful to I.V. Gladyshev, who carried out the corresponding numerical calculations) and experiments conducted with diamond of type I (stars) and with diamond of type II (triangles) (see the above-mentioned reference [138].

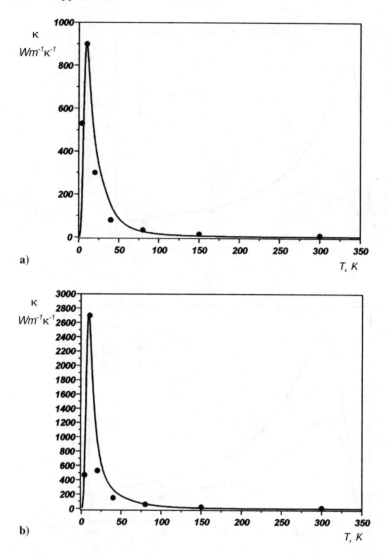

Fig. B.2. Theoretical temperature dependence of the thermal conductivity coefficient for glass (**a**) SiO$_2$ and (**b**) crystal NaCl. The points are the experimental data from [138].

1. In the above statement of the problem the role of surface phonons in the theory of internal microscopic relaxation in crystal dielectrics has been clarified.

2. A detailed relaxation theory has been presented, in which the four main interacting subsystems were taken into account, namely, the longitudinal, transverse, optical and surface phonons of the thermostat.

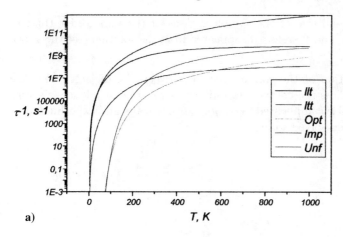

a)

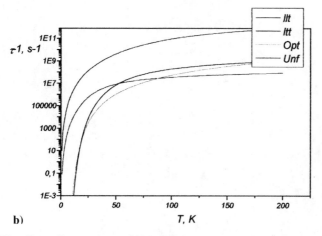

b)

Fig. B.3. Comparison of the different reverse time relaxation curves for (a) diamond and (b) NaCl calculated by the numerical integration method

3. As a result of the relaxation theory, the coefficient of thermal the conductivity has been presented in the most general form taking account of most of the important interaction mechanisms.

4. By means of numerical integration precise theoretical curves of the dependence $\kappa(T)$ (for two types of crystal dielectrics: diamond and sodium salt NaCl) were constructed which perfectly correspond to the experimental points (see Fig. B.1)

5. In drawing the curve $\kappa(T)$ only the experimentally found parameters were plotted, and of great significance is the fact that we did not use any adjusted parameter, besides, maybe, in the contact region δ, which,

due to the small size of the sample L, was of the same order (but, of course, lower), and therefore disappeared from the solution (see the above estimations).

The investigation of the contact region between the dielectric and the thermostat as well as its (contact region!) temperature dependence is a very important and quite complicated task, both theoretically and experimentally, which has not been solved so far.

Appendix C: Simplified Theory of Thermal Conduction in a Porous Dielectric

The theoretical description of the thermal conduction process even in the simplest crystal structures (as we already know from the above text) is connected with rather awkward equations requiring us to account for the symmetry properties of the crystal lattice and to be extremely careful in analyzing the microscopic processes of internal energy dissipation in the class of substances under investigation [144–149] (see also Appendix B).

The number of relaxation mechanisms increases nonlinearly, depending on the type of internal physical structure of the sample under investigation, and the same substances may be arranged as a hierarchical series according to increasing complexity: dielectrics; magnetics; metals; ferromagnetic metals; semiconductors; alloys; composites; porous structures.

Incidentally, there always exists a real chance to miss some effective dissipative mechanisms of interaction playing a very important role in the true character of thermodynamic equilibrium establishment. Indeed, if we omit them, it will radically change (to the wrong side!) the qualitative course of thermal balance establishment and in future will lead to wrong conclusions concerning the functional (for example, thermal) behavior of some macroscopic dissipative characteristics (say, the conductivity or thermal conductance) of the substance under study.

It should be noted that it is the *correct* theory of relaxation, which allows us to develop the theory of magnetic susceptibility in magnetic materials and to find out the *correct* temperature dependences of the tensor components of the magnetic susceptibility χ_{ik} (see, for instance, the review [83] and the paper [150]).

As one of the most complicated classes of structures porous, crystal dielectrics are very interesting objects in terms of the analytical study of the thermo-physical characteristics of such substances, and our present choice of this crystal porous dielectric (as an example to demonstrate the description method of thermal conduction in complex inhomogeneous structures) is not arbitrary. Firstly, it is connected with the minimum set of relaxation mechanisms (defining the dependence $\kappa(T, L, R, \lambda, m)$, where, as we remember, the parameter $\lambda = c_l/c_t$ and λ is always larger than one, and c_l and c_t are the longitudinal and transverse sound velocities, respectively) as compared with any other type of crystal substance. Secondly, it is connected with a clearly understood relaxation pattern occurring in a pure (homogeneous in composition) dielectric, whose qualitative description is given in Sect. 4.7, and a very detailed quantitative description is given in the Appendix B.

Based on the above analysis, we shall as maximally as possible simplify all the calculations and show how to estimate $\kappa(T, L, R, \lambda, m)$ for substances of inhomogeneous composition from the example of a porous dielectric.

Thus, in order to find the analytical dependence of $\kappa(T, L, R, \lambda, m)$, we shall use the general expression for the thermal conductivity coefficient of an inhomogeneous medium (see monographs [86, 126]) and present it in the following form convenient for analysis (cf. Chap. 6):

$$\kappa = \kappa_{11}^* + \kappa_{12}^* + \kappa_{21}^* + \kappa_{22}^*, \tag{C.1}$$

where the components are

$$\begin{cases} \kappa_{11}^* = (1-m)^2 \kappa_{11}, \\ \kappa_{12}^* = m(1-m)\left(\dfrac{h_{12}}{R}\right)\kappa_{12}, \\ \kappa_{21}^* = m(1-m)\left(\dfrac{h_{21}}{R}\right)\kappa_{21}, \\ \kappa_{22}^* = m^2 \kappa_{22}. \end{cases} \tag{C.2}$$

Here κ_{11} is the thermal conductivity coefficient of a pure dielectric whose estimational equation is given below (see formula (C.3)), κ_{22} is the thermal conductivity coefficient of the substance filling the pores, κ_{12} and κ_{21} are the same coefficients but for the contact region ("transition" layer) between the main matrix (phase "1") and a pore (phase "2"), respectively, for the phonon–photon (index "12") and photon–phonon (index "21") mechanisms of interaction (according to Sect. 4.7). The parameters $h_{12} \times h_{21}$ correspond to the linear dimension of the transition region and can be taken as equal in order of magnitude to the free path of a phonon, i.e. $h_{12} \approx l_{12} \approx c_s \tau_{12}$, and a photon, i.e. $h_{21} \approx l_{21} \approx [c/(\varepsilon\mu)^{1/2}]\tau_{21}$, where c_s is the average sound velocity in the dielectric, c is the speed of light in a vacuum, ε is the dielectric permittivity and μ is the magnetic permeability of the main matrix.

In order to estimate κ_{11}^* we shall apply the gas-kinetic approach and hold that

$$\kappa_{11}^* = (1-m)^2 \kappa_{11} = (1-m)^2 \left[\frac{\hbar\nu c_l^2}{2\pi^2}\right]\int k^2 \omega_{lk}\tau_{lk}^*\left(\frac{\partial \bar{N}_{lk}}{\partial T}\right)\mathrm{d}k, \tag{C.3}$$

where the parameter $\nu = 1/[3^{1/3}(1+\lambda^3)]^{2/3}$ (for more information see Appendix B), $\omega_{lk} = c_l k$, k is the wave vector and N_{lk} is the equilibrium Bose distribution function of the longitudinal phonons. The inverse effective relaxation time appearing in (C.3) is (according to the rule of composition of probabilities of incompatible events) just the sum of the inverse times:

$$\frac{1}{\tau_{lk}^*} = \frac{c_l}{L} + \frac{1}{\tau_{lltk}} + \frac{1}{\tau_{lttk}} + \frac{1}{\tau_{ltimpk}} + \frac{1}{\tau_{lOk}}. \tag{C.4}$$

Determination of the relaxation times contained in (C.4) can be found in the formulae (B.25)–(B.27) and (B.29).

Let us accept on a purely phenomenological basis that the factorized relation

$$\kappa_{12}^* = m(1-m)\left(\frac{h_{12}}{R}\right)\kappa_{12}$$

may also be presented in the gas-kinetic approximation and written in the form of the following integral:

$$\kappa_{12}^* = \nu m(1-m)c_l^3 \frac{\hbar}{2\pi^2 R} \int\limits_0^{\pi/a} k^2 \omega_{lk}\tau_{l-\text{phot}k}^{*2}\left(\frac{\partial N_{lk}}{\partial T}\right)dk, \tag{C.5}$$

where the phonon–photon relaxation time, denoted as $\tau_{l-\text{phot}k}^*$, is determined by the formula:

$$\tau_{l-\text{phot}k}^{*-1} = \frac{c_l}{L + \tau_{l-\text{phot}k}^{-1}}. \tag{C.6}$$

Note, by the way, that in terms of the indices "1" and "2" the time $\tau_{l-\text{phot}k}^*$ should be written as τ_{12k}. The calculation of $\tau_{l-\text{phot}k}$ is performed below and is given by (C.32).

By analogy to the equation (C.5), also purely phenomenologically, let us take

$$\kappa_{21}^* = m(1-m)c^3 \frac{\hbar}{6\pi^2 R}(\varepsilon\mu)^{1/2} \int\limits_0^{\pi/a} q^2 \omega\tau_{\text{phot}-lq}^{*2}\left(\frac{\partial f_\omega}{\partial T}\right)dq, \tag{C.7}$$

where the photon dispersion in the medium is $\omega = kc/(\varepsilon\mu)^{1/2}$ and the equilibrium distribution function f_ω is the Planck function, i.e. $f_\omega = [\exp(\hbar\omega/T) - 1]^{-1}$.

Here the relaxation time is determined by the equation:

$$\tau_{\text{phot}-lk}^{*-1} = \frac{c}{L(\varepsilon\mu)^{1/2}} + \tau_{\text{phot}-lk}^{-1}. \tag{C.8}$$

As well as the time $\tau_{l-\text{phot}}$, the average time (over the equilibrium distribution function of photons) of the photon relaxation time $\tau_{\text{phot}-l}$ will be calculated later (formula (C.33)).

It should be also noted that any average value of relaxation inverse time τ^{-1} may be found by averaging the inverse time τ_k^{-1} over the equilibrium Bose function f_k according to the rule: $\tau^{-1} = \int f_k \tau_k^{-1} k^2 dk / \int f_k k^2 dk$.

Finally, the last term in (C.1) for the mechanism of radiant heat transfer in the gas-kinetic approximation may be presented as

$$\kappa_{22}^* = m^2\kappa_{22} = m^2\left[\frac{\hbar c^2}{6\pi^2\varepsilon\mu}\right]\int\limits_0^\infty q^2 \omega\tau_{\text{phot}-lq}^*\left(\frac{\partial f_q}{\partial T}\right)dq. \tag{C.9}$$

The first terms in the formulae (C.6) and (C.8) are connected with the so-called Knudsen mechanism of phonon thermalization at the sample boundary

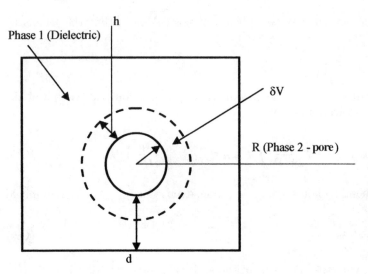

Fig. C.1. The bubble is a free volume (pore) with radius R. The parameters are related as follows: $h \sim R < d$ (see Appendix C)

(see C.6) (and the discussion in Sect. 4.7) and of the analogous effect for a photon (see (C.8)). As the temperature decreases the probabilities $\tau^{-1}_{\text{phot}-l}$ and $\tau^{-1}_{l-\text{phot}}$ tend to zero, and the last stage of relaxation is completed at the boundary of the dielectric due to the connection with the thermostat.

It should be stressed also that the dimensionless parameters $c_l \tau^*_{l-\text{phot}}/R$ and $c\tau^*_{\text{phot}-l}/R(\varepsilon\mu)^{1/2}$ in (C.5) and (C.7) are not at all assumed to be small! It is related to the fact that the width in the region of contact h_{12} or h_{21} on the interface boundary may exceed the free volume radius R, remaining, however, small compared to the distance d from the free volume surface to the dielectric boundary (see Fig. C.1).

In order to estimate the relaxation times $\tau_{\text{phot}-l}$ and $\tau_{l-\text{phot}}$ (in other words, the times τ_{12} and τ_{21}), we shall suppose that the crystal structure of the dielectric main matrix has cubic symmetry and may be referred to the symmetry group of, say, zinc blende T_d. Then, if we introduce the coupling constant of the electrostriction mechanism G, we may present the invariant Hamiltonian interaction for the elastic and electromagnetic waves in the following form:

$$H_{\text{int}} = G \int_{\delta V} (E^2 \text{div}\, \boldsymbol{u} + 2u_{ik}E_i E_k)\mathrm{d}^3 x \,. \tag{C.10}$$

We did not introduce "caps" over the corresponding magnitudes, for we are sure that it will not mislead the reader, but for us it will significantly simplify recording.

The indices i, k relate to the Cartesian coordinates x, y, z and the repeated indices traditionally imply summation. The integration region δV in (C.10) is the volume in the vicinity of a pore and for a spherical cavity of radius R it can be found from the formula:

$$\delta V = 4\pi R^2 h \,, \tag{C.11}$$

where h is the extinction coefficient, or in other words, the depth of penetration of non-equilibrium electromagnetic radiation from a pore into the main matrix, relating, according to the law of Boggier–Lambert, to the region of absorption of photons by a substance. In other words, h is some average photon free path in a medium up to its complete annihilation due to the interaction with density fluctuations (phonons). The interaction between the radiation and the dielectric atoms is insignificant, for it is assumed that the photon wavelength for the black-body radiation λ is considerably larger than the linear dimension of an atom a, i.e. $\lambda \gg a$. Thus, the extinction coefficient is

$$h \approx \left[\frac{c}{(\varepsilon\mu)^{1/2}} \right] \tau_{l-\text{phot}} \,. \tag{C.12}$$

Before we proceed to further formalism, we should note here that though the laws of conservation of energy and momentum allow processes with the participation of one photon and, for example, of one longitudinal and one transverse phonon, however, the interaction of the type corresponding to $G^*_{ik/nm} \int E_i u_{kl} u_{nm} \mathrm{d}^3 x$ is forbidden by the symmetry rules of selection.

Indeed, for an ordinary cubic lattice there is the element of the coordinate transformation group of the type $x \to -x$, $y \to y$, $z \to z$, and hence there follows the conclusion that all the coefficients $G^*_{ik/nm}$ become identically zero.

Further, in (C.10) the deformation tensor is known from Appendix B and is given by the relation (A2.6), and the electric field operator is introduced according to the definition:

$$\boldsymbol{E} = -c^{-1} \frac{\partial \boldsymbol{A}}{\partial t} \,, \tag{C.13}$$

where $\boldsymbol{A}$ is the vector wave function of a free photon.

Let us introduce the following second quantized operators:

$$A_1 = \sum_{q,\alpha} \left(\frac{2\pi\hbar c}{qV_2} \right)^{1/2} e_{i\alpha}(c^+_{q\alpha} e^{-i\omega t} + c_{-q\alpha} e^{i\omega t}) e^{iqx} \tag{C.14}$$

and

$$u_i = \sum_{k,\beta,j} e_i \left[\frac{\hbar}{2\rho\omega_j(V - V_2)} \right]^{1/2}$$
$$\times [b^+_{k\beta} \exp(-i\omega_j t) + b_{k\beta} \exp(i\omega_j t)] e^{ikx} \,, \tag{C.15}$$

where $c_{q,\alpha}^+ (c_{q,\alpha})$ is the operator of creation (annihilation) of a photon with wave vector q and polarization α, $b^+(b)$ is the same for a phonon and $V_2 = (4\pi/3)R^3$ is the volume of a pore.

According to the definition of the electric field operator (C.13) we have

$$\boldsymbol{E} = \mathrm{i} \left(\frac{2\pi\hbar}{cV_2}\right)^{1/2} \sum_{q,\alpha} \boldsymbol{e}_\alpha \left(\frac{\omega}{k^{1/2}}\right) [c_{q,\alpha}^+(t) - c_{q,\alpha}(t)]\mathrm{e}^{\mathrm{i}qx}, \qquad (C.16)$$

where for brevity we have introduced the denotation of the photon creation (annihilation) operator in the representation of interaction (Chap. 5), i.e.

$$\begin{cases} c_{q,\alpha}^+(t) = c_{q,\alpha}^+ e^{-\mathrm{i}\omega t} \quad \text{and} \\ c_{q,\alpha}(t) = c_{q,\alpha} e^{\mathrm{i}\omega t}. \end{cases} \qquad (C.17)$$

By close analogy to (C.16) the operator of the continuum deformation tensor is obtained (see Appendix B, formulae (A2.2) and (A2.6)), which, if there is a pore, should be presented as:

$$u_{is} = 0.5\mathrm{i} \sum_{k,\beta,j} (e_i k_s + e_s k_i) \left[\frac{\hbar}{2\rho\omega_j(V - V_2)}\right]^{1/2}$$
$$\times [b_{k\beta}^+ \exp(-\mathrm{i}\omega_j t) + b_{k\beta} \exp(\mathrm{i}\omega_j t)]\mathrm{e}^{\mathrm{i}kx}, \qquad (C.18)$$

For familiarization purpose we shall submit to the reader the equation for the operator of magnetic field intensity, which, using the definition $\boldsymbol{H} = \mathrm{rot}\,\boldsymbol{A}$ and the equation (C.14), may be introduced in the form:

$$\boldsymbol{H} = \mathrm{i} \sum_{q,\alpha} [\boldsymbol{e}_\alpha \times \boldsymbol{q}] \left(\frac{2\pi\hbar c}{qV_2}\right)^{1/2} [c_{q\alpha}^+(t) + c_{-q\alpha}(t)]\mathrm{e}^{\mathrm{i}qx}. \qquad (C.19)$$

Using (C.16), (C.18) and (C.17) the Hamiltonian (C.10) is as follows:

$$H_{\mathrm{int}} = \sum_{q,q',k} \sum_{\{\alpha\}} \sum_{\{\beta\}} \psi_{\{\alpha,\beta\}}(\boldsymbol{q},\boldsymbol{q}',\boldsymbol{k})(c_{q,\alpha}^+ - c_{q,\alpha})(c_{q',\alpha'}^+ - c_{q',\alpha'})(b_{k\beta}^+ + b_{k\beta})$$
$$\times \Delta(\boldsymbol{q} + \boldsymbol{q}' + \boldsymbol{k}), \qquad (C.20)$$

where the process amplitude is

$$\psi_{\{\alpha,\beta\}}(\boldsymbol{q},\boldsymbol{q}',\boldsymbol{k}) = \mathrm{i} \left(\frac{\delta V}{V_2}\right) \left[\frac{2\pi\hbar cG}{(\varepsilon\mu)^{1/2}}\right] \left[\frac{\hbar}{2\rho(V - V_2)}\right]^{1/2} \left(\frac{qq'}{\omega_{lk}}\right)^{1/2} \qquad (C.21)$$
$$\times [(\boldsymbol{e}_\alpha \boldsymbol{e}_{\alpha'})(\boldsymbol{e}_\beta \boldsymbol{k}) + (\boldsymbol{e}_\alpha \boldsymbol{e}_\beta)(\boldsymbol{e}_{\alpha'}\boldsymbol{k}) + (\boldsymbol{e}_{\alpha'}\boldsymbol{e}_\beta)(\boldsymbol{e}_\alpha \boldsymbol{k})].$$

The collision integrals are obtained from the Hamiltonian (C.20) by the algorithm which we know from Appendix B. Indeed, for phonon–photon scattering we have

$$L_{l-\text{phot}}\{N_{lk}\} = \left(\frac{2\pi}{\hbar^2}\right) \sum_{q,q'} \sum_{\alpha\alpha'} |\psi_{\{\alpha,\beta\}}(\boldsymbol{q}, \boldsymbol{q}', \boldsymbol{k})|^2$$

$$\times [(1 + N_{lk})(1 + f_q)f_{q'} - N_{lk}f_q(1 + f_{q'})]$$

$$\times \Delta(\boldsymbol{k} + \boldsymbol{q} - \boldsymbol{q}')\delta\left[\omega_{lk} + \left(\frac{c}{(\varepsilon\mu)^{1/2}}\right)(q - q')\right]. \qquad (C.22)$$

The second collision integral for the photon-phonon interaction mechanism will be obviously the following:

$$L_{\text{phot}-l}\{f_q\} = \left(\frac{2\pi}{\hbar^2}\right) \sum_{q',k} \sum_{\alpha,\beta} |\psi_{\{\alpha,\beta\}}(\boldsymbol{q}, \boldsymbol{q}', \boldsymbol{k})|^2$$

$$\times \{[(1 + f_q)f_{q'}(1 + N_{lk}) - f_q(1 + f_{q'})N_{lk}]$$

$$\times \Delta(\boldsymbol{k} + \boldsymbol{q} - \boldsymbol{q}')\delta\left[\omega_{lk} + \left(\frac{c}{(\varepsilon\mu)^{1/2}}\right)(q - q')\right]$$

$$+ [(1 + f_q)f_{q'}N_k - f_q(1 + f_{q'})(1 + N_k)]$$

$$\times \Delta(\boldsymbol{q} - \boldsymbol{q}' - \boldsymbol{k})\delta\left[\frac{\omega_{lk} - c}{(\varepsilon\mu)^{1/2}}(q - q')\right]\}. \qquad (C.23)$$

By means of the given collision integrals we may estimate the relaxation times. According to the definition (B.23) we have

$$\begin{cases} \dfrac{1}{\tau_{l-\text{phot}k}} = -\left.\dfrac{\delta L_{l-\text{phot}}\{N_{lk}\}}{\delta N_{lk}}\right|_{N_{lk}=\bar{N}_{lk}}, & (C.24) \\[3mm] \dfrac{1}{\tau_{\text{phot}-lk}} = -\left.\dfrac{\delta L_{\text{phot}-l}\{f_q\}}{\delta f_q}\right|_{f_q=\bar{f}_q}, & (C.25) \end{cases}$$

If we now omit the rather simple calculations with which the reader is already acquainted, we shall find the following two analytical equations:

$$\begin{cases} \dfrac{1}{\tau_{l-\text{phot}k}} = N\left(\dfrac{2\pi}{\hbar^2}\right) \sum_{q,q'} \sum_{\alpha\alpha'} |\psi_{\{\alpha,\beta\}}(\boldsymbol{q}, \boldsymbol{q}', \boldsymbol{k})|^2 (\bar{f}_q - \bar{f}_q) \\[3mm] \qquad\qquad \times \Delta(\boldsymbol{k} + \boldsymbol{q} - \boldsymbol{q}')\delta\left[\omega_{lk} + \left(\dfrac{c}{(\varepsilon\mu)^{1/2}}\right)(q - q')\right], \qquad (C.26) \\[4mm] \dfrac{1}{\tau_{\text{phot}-lq}} = N_{\text{phot}}\left(\dfrac{2\pi}{\hbar^2}\right) \sum_{k,q'} \sum_{\alpha\alpha'} |\psi_{\{\alpha,\beta\}}(\boldsymbol{q}, \boldsymbol{q}', \boldsymbol{k})|^2 \qquad\qquad (C.27) \\[3mm] \qquad \times \left\{(\bar{N}_{lk} - \bar{f}_{q'})\Delta(\boldsymbol{k} + \boldsymbol{q} - \boldsymbol{q}')\delta\left[\omega_{lk} + \left(\dfrac{c}{(\varepsilon\mu)^{1/2}}\right)(q - q')\right]\right. \\[3mm] \qquad \left. + (1 + \bar{N}_{lk} + \bar{f}_{q'})\Delta(\boldsymbol{k} + \boldsymbol{q}' - \boldsymbol{q})\delta\left[\omega_{lk} + \left(\dfrac{c}{(\varepsilon\mu)^{1/2}}\right)(q' - q)\right]\right\}, \end{cases}$$

where N is the number of atoms in the main dielectric matrix, i.e. in the volume $V - V_2$, and N_{phot} is the number of photons in the volume V_2.

In order to perform further calculations, in the formulae (C.26)–(C.28) we should pass from summation over "q", "q'" and "k" to integration over the corresponding variables. This procedure incidentally is carried out by means of the substitutions:

$$\sum_k (\ldots) = (V - V_2) \int (\ldots) \frac{\mathrm{d}^3 k}{(2\pi)^3}, \tag{C.28}$$

$$\sum_q (\ldots) = V_2 \int (\ldots) \frac{\mathrm{d}^3 q}{(2\pi)^3}. \tag{C.29}$$

From the analysis of the energy and momentum conservation laws for the interaction mechanism "l-phot" (equation (C.26)) it follows that the virtual photon vector (over which the summation is performed!) must be in the interval (we recommend the reader to carry out independently the corresponding simple algebraic transformations):

$$0.5(1 - b)k \leq q' < \infty, \tag{C.30}$$

where the parameter $b = c_l(\varepsilon\mu)^{1/2}/c \ll 1$.

Analogously, it is easy to show that for the first mechanism in (C.28) the following inequality should hold:

$$q \leq q' \leq \frac{(1 + b)q}{(1 - b)}, \tag{C.31}$$

and for the second one

$$\frac{q(1 - b)}{(1 + b)} \leq q' \leq q. \tag{C.32}$$

Thus, taking account of the performed analysis of the conservation laws and the obtained equations for the virtual wave vectors, we obtain the following two remarkable formulae:

$$\frac{1}{\tau_{l-\text{phot}}}(y) = \left(\frac{b^2}{\tau_0}\right) \int_0^\infty [(x + 0.5y)^2 - 0.25b^2 y^2]^2$$
$$\times [N(x + 0.5(1 - b)y) - N(x + 0.5(1 + b)y)]\mathrm{d}x, \tag{C.33}$$

$$\frac{1}{\tau_{\text{phot}-l}}(y) = \left(\frac{1}{\tau_0}\right) \left\{ \int_y^{x'(y)} (x - y)^2 x^2 [N(x - y) - N(x)]\mathrm{d}x \right.$$
$$\left. + \int_{x''(y)}^y (y - x)^2 x^2 [1 + N(x) + N(y - x)]\mathrm{d}x \right\}, \tag{C.34}$$

where $x'(y) = y(1 + b)/(1 - b)$, $x''(y) = y(1 - b)/(1 + b)$, and the inverse time τ_0 is introduced by the equation:

$$\frac{1}{\tau_0} = \left(\frac{3\pi^6}{25}\right) G^2 b^2 \left(\frac{\delta V^2}{V_2}\right) \left(\frac{\hbar N_{\text{phot}}}{Ma^5}\right) \left(\frac{c_s}{c_l}\right)^5 \left(\frac{T}{\theta_D}\right)^5, \tag{C.35}$$

where M denotes the atomic mass and $M = \rho a^3$, and the Debye temperature is introduced according to the definition $\theta_D = \pi \hbar c_s/a$. The average sound velocity is $c_s = 3^{1/3} c_l/(1 + 2\lambda^3)^{1/3}$. The function $N(x)$ is the ordinary Bose distribution function having the form $N(x) = (e^x - 1)^{-1}$. We have omitted the "bar" over all the equilibrium distribution functions for the sake of simplification.

So far as the dimensionless integration variables in (C.33) and (C.34) are concerned, they were introduced by the following equations: for (C.33) by $y = \hbar c_l k/T$, and for (C.34) by $y = \hbar c q/T$.

The dependence (C.35) determines the relation with one of the average relaxation times: either with $\tau_{l-\text{phot}}$, or with $c\tau_{\text{phot}-l}$. Let us explain this point in more detail.

According to the dependences (C.11) and (C.12) we have either

$$\begin{cases} \delta V = 4\pi R^2 h = 4\pi R^2 c_s \tau_{l-\text{phot}}, & \text{or} \\ \delta V = 4\pi R^2 h = 4\pi R^2 c\tau_{\text{phot}-l}. \end{cases} \tag{C.36}$$

Which of the given relations should be chosen is decided by simple estimation. If, say, the product $c_s \tau_{l-\text{phot}}$ is smaller than the product $c\tau_{\text{phot}-l}$, then it is the former which is left, for the smaller product should appear in κ. The relation between them will be dealt with later. The number of photons N_{phot} are found in the following way:

$$N_{\text{phot}} = V_2 \int\limits_0^{q_{max}} \frac{q^2 \mathrm{d}q}{2\pi^2}.$$

Ultimately we have

$$\frac{N_{\text{phot}}}{V_2} = \frac{q_{\text{max}}^3}{6\pi^2}, \tag{C.37}$$

where q_{max} is some limiting wave vector, defined by the upper boundary of the optical range (or lower boundary of X-rays). Hence it is clear that

$$q_{\text{max}} \approx 10^5 (\text{cm}^{-1}).$$

The true relaxation times are given by (C.33) and (C.34), which, in turn, are to be averaged with respect to the equilibrium Bose distribution function $N(y)$. As a result we obtain the two dimensionless functions:

$$U(b,T) = \left(\frac{\tau_0}{\tau_{l-\text{phot}}}\right) = \left(\frac{b^2}{g_1}\right) \int\limits_0^{\theta_D^*/T} N(y) y^2 \mathrm{d}y \int\limits_0^{\infty} [(x + 0.5y)^2 - 0.25b^2 y^2]$$

$$\times \{N[(x + 0.5(1-b)y] - N[(x + 0.5(1+b)y]\} \mathrm{d}x \tag{C.38\,a}$$

and

$$V(b,T) = \left(\frac{\tau_0}{\tau_{\text{phot}-l}}\right) = \left(\frac{b^2}{g_2}\right)\left\{\int_0^\infty N(y)y^2\mathrm{d}y \int_y^{x'(y)} x^2(x-y)^2\right.$$

$$\times[N(x-y)-N(x)]\mathrm{d}x + \int_0^\infty N(y)y^2\mathrm{d}y \int_{x''(y)}^{y} x^2(y-x)^2$$

$$\left. \times[1+N(x)-N(y-x)]\mathrm{d}x\right\}, \qquad (\text{C.38 b})$$

where

$$g_1 = g_1(T) = \int_0^{\theta_D^*/T} N(y)y^2\mathrm{d}y\,,$$

and

$$g_2 = \int_0^\infty N(y)y^2\mathrm{d}y \approx 2.4\,.$$

It is quite clear that the precise relation between the functions $U(b,T)$ and $V(b,T)$ may be found only by means of numerical integration of the equations (C.38 a)–(C.38 b). The performed analysis (see Figs. C.2a,b) showed that $U(b,T) \gg V(b,T)$ throughout the whole real range of the change of the temperature and the parameter b, and, therefore, the time $\tau_{l-\text{phot}}$ is considerably less than the time $\tau_{\text{phot}-l}$.

That is why the coefficient of thermal conductivity and its parametrical dependences, according to the formula (C.1) must be defined by the relation

$$\kappa = \kappa_{11}^* + \kappa_{12}^* + \kappa_{22}^*\,, \qquad (\text{C.39})$$

where, in accordance with the above reasoning, only the coefficient κ_{12}^* is left.

The formula (C.39) actually completes the analysis of the thermal conduction process in a crystal dielectric with spherical pores of radius R.

The complete analytical expression for κ taking account of the equations (C.2), (C.3), (C.5) and (C.9) may be presented in a very awkward and quite immense form.

Indeed, collecting these formulae together, we obtain the following for an inhomogeneous substance:

$$\kappa = (1-m)^2\left[\frac{\hbar\nu c_l^2}{2\pi^2}\right]\int_0^{\pi/a} k^2\omega_{lk}\tau_{lk}^*\left(\frac{\partial N_{lk}}{\partial T}\right)\mathrm{d}k$$

$$+ \nu m(1-m)c_l^3\frac{h}{2\pi^2 R}\int_0^{\pi/a} k^2\omega_{lk}\tau_{l-\text{photk}}^{*2}\left(\frac{\partial N_{lk}}{\partial T}\right)\mathrm{d}k$$

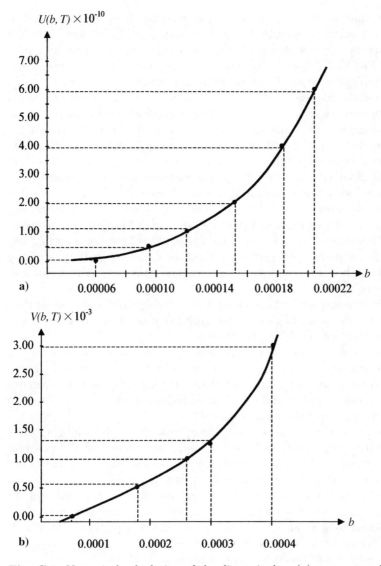

Fig. C.2. Numerical calculation of the dimensionless (**a**) parameters $U(b,T)$ (**b**) and $V(b,T)$

$$+ m^2 \left[\frac{\hbar c^2}{6\pi^2 \varepsilon \mu} \right] \int_0^\infty q^2 \omega \tau^*_{\text{phot}-lq} \left(\frac{\partial f_q}{\partial T} \right) dq \,, \tag{C.40}$$

where the effective relaxation times are defined by (C.4), (C.6) and (C.8).

Further analysis of the relatively simple formula (C.40) (as compared to the rather complicated formula obtained in Chap. 5; see (5.88) and (5.90)), is possible only by means of the application of exclusively numerical methods.

Concluding this Appendix we would like to note that our purpose has been achieved and we have shown by the final equation (C.40) how estimation equations for the thermal conductivity coefficients of inhomogeneous media may be obtained relatively quickly and rather effectively by means of the simple gas-kinetic approximation. And although we have considered only the case of a crystal dielectric with only a single pore, the general algorithm and the submitted approach do, however, seem to allow its application (though in a somewhat modified form) for thermal conductivity analysis in considerably more complex structures containing a whole ensemble of pores and other inhomogeneities rather than one pore. The only requirement necessary for the application of this method is to meet the condition that the distances between inhomogeneities d should *exceed* (we stress: not considerably but just exceed!) the double penetration depth of the non-equilibrium black-body radiation into the transition zone δV. Mathematically it is written down as the inequality $d > 2h$.

Note, by the way, that this condition is necessary and at the same time sufficient from the point of view of the application of the above calculations and the general formulae (C.1)–(C.2).

A curious reader may ask a reasonable question: why was it necessary in Chap. 5 to carry out calculations using the rather complicated mathematical apparatus involving Kubo's formula, the Green's function, the non-equilibrium density matrix and the diagram technique, while κ may be quite easily estimated from the equation (C.40)?

The reason is that besides familiarizing character this monograph has the instructive purpose of teaching the beginning researcher the powerful methods of theoretical physics applied nowadays in science. But that is not all! Formulae (C.1)–(C.2), if examined attentively, were introduced purely *phenomenologically* with the porosity already included (!). Equations (5.88) and the following ones were obtained rigorously, and the porosity arose there as a *mathematically valid* and proved result. In the equations which were accepted without proof in the form of equations (C.1)–(C.2) the dependence on the porosity m was based on the general formula (5.90).

Thus, due to the method of the theory of non-equilibrium processes and in particular due to the Kubo formula, which was modified in applying it to our problem, it turned out to be possible to justify rigorously the introduction of the porosity m. The method of the quasi-classical Boltzmann equation could not cope with this task and hence the porosity would have to be introduced "with hands", and a priori the functional dependence of κ on m would not be known!! Using the method of the quasi-classical Boltzmann equation it is impossible to introduce the porosity m, and, moreover, a priori it is not known what the functional dependence of κ on m should be!!

Appendix D

D.1 Theory of Fluctuations

Some data on the theory of fluctuations were touched upon in Sect. 6.3, and now we shall deal with this theoretically very important subject and shall dwell upon some peculiarities of its physical and mathematical character in more detail.

We know (see [10]) that the entropy is proportional to the probability of realization of some event x (x should be understood as any thermodynamic independent parameter: temperature, pressure, the number of particles, entropy, etc.), i.e.

$$S(x) = \tilde{c}\ln w(x).$$ (D.1)

Hence the desired probability is

$$W(x) = c\,e^{S(x)}.$$ (D.2)

In the equilibrium state the system entropy is a maximum and therefore the following two inequalities are valid:

$$\frac{\partial S}{\partial x} = 0, \quad \frac{\partial^2 S}{\partial X} < 0.$$ (D.3)

Consequently, in the immediate vicinity of the equilibrium point it would be justified to present (D.2) as an expansion into a Taylor series:

$$W(x) = c\,e^{S(\bar{x}) - \frac{\alpha}{2}(x-\bar{x})^2},$$ (D.4)

where x is the solution of the equation $\partial S/\partial x = 0$. Redenoting the variables in (D.4), i.e., taking

$$c \Rightarrow c\,e^{S(x)},$$

we find

$$W(x) = c\,e^{-\frac{\alpha}{2}(x-\bar{x})^2}.$$ (D.5)

Since the integral of $W(x)$ over x in the limits $-\infty$ and $+\infty$ is a real event, from the given condition the constant c may be determined. Indeed, the normalization condition gives

$$c \int_{-\infty}^{+\infty} \exp\left(-\frac{\alpha(x-x)^2}{2}\right) dx = 1 \,.$$

Hence, as the Poisson integral is

$$J = \int_{-\infty}^{\infty} e^{-\frac{\alpha}{2}(x-\bar{x})} dx = \sqrt{\frac{2\pi}{\alpha}} \,,$$

we have $c = \sqrt{\dfrac{\alpha}{2\pi}}$, and therefore

$$W(x) = \sqrt{\frac{\alpha}{2\pi}} e^{-\frac{\alpha}{2}(x-\bar{x})^2} \,. \tag{D.6}$$

Equation (D.6) is nothing but Gauss's distribution or, as it is named in probability theory, the normal distribution. By means of (D.6) we shall calculate the mean square value of the fluctuation of the random quantity x. We have

$$\overline{x^2} = \sqrt{\frac{\alpha}{2\pi}} \int_{-\infty}^{\infty} x^2 e^{-\alpha\frac{(x-\bar{x})^2}{2}} \, dx = \left(x - \bar{x} = \sqrt{\frac{2}{\alpha}}\xi\right)$$

$$= \sqrt{\frac{\alpha}{2\pi}} \int_{-\infty}^{\infty} \left(\bar{x} + \sqrt{\frac{2}{\alpha}}\xi\right)^2 e^{-\xi^2} \sqrt{\frac{2}{\alpha}} d\xi$$

$$= \bar{x}^2 \sqrt{\frac{1}{\pi}} \int_{-\infty}^{\infty} e^{-\xi^2} d\xi + \frac{2}{\alpha\sqrt{\pi}} \int_{-\infty}^{\infty} \xi^2 e^{-\xi^2} d\xi = \bar{x}^2 + \frac{1}{\alpha} \,.$$

Thus, we get

$$\frac{1}{\alpha} = \overline{x^2} - \bar{x}^2 \,. \tag{D.7}$$

Let us introduce traditional symbolism:

$$\langle \delta x^2 \rangle = \frac{1}{\alpha} = \overline{x^2} - \bar{x}^2 \,. \tag{D.8}$$

Then the distribution (D.6) becomes equal to

$$W(x) = \frac{1}{\sqrt{2\pi \langle \delta x^2 \rangle}} e^{-\frac{(x-\bar{x})^2}{2\langle \delta x^2 \rangle}} \,. \tag{D.9}$$

The value $\langle \delta x^2 \rangle$ is called the fluctuation of the random quantity x. In the theory of probability the fluctuation $\langle \delta x^2 \rangle^{1/2}$ is conventionally denoted as σ, which is called the root-mean-square deviation.

D.2 Fluctuations of the Main Thermodynamic Values

The calculation of the fluctuations of the main thermodynamic values actually is confined to the application of the general formula (D.9). The only fine point here is connected with the choice of the variables whose fluctuations are to be calculated. We shall show from concrete examples how formula (D.9) operates, but first we should return to equation (D.2), expressing it by means of the fluctuation entropy change ΔS. We have

$$W \sim e^{\Delta S}. \tag{D.10}$$

In terms of the variables S and V the energy change is

$$\Delta E = T_0 \Delta S - P_0 \Delta V, \tag{D.11}$$

where T_0 and P_0 are the equilibrium values of the temperature and pressure. On the other hand, ΔS must be equal to $-R_{\min}/T_0$, where $R_{\min}$ is the minimum work which is to be done in order to cause in a *reversible* way the assigned change of thermodynamic parameters. It means that in terms of the variables S and V we should take

$$R_{\min} = \Delta E - T_0 \Delta S + P_0 \Delta V, \tag{D.12}$$

Further, we shall omit the subscripts "0" on T and P which imply equilibrium values.

In terms of the variables T and V, we have

$$R_{\min} = (\Delta F + S\Delta T + P\Delta V). \tag{D.13}$$

In terms of the variables P and T, we have

$$R_{\min} = -(\Delta \Phi + S\Delta T - V\Delta P). \tag{D.14}$$

In terms of the variables S and P, we have

$$R_{\min} = \Delta W - T\Delta S - V\Delta P. \tag{D.15}$$

As an example we choose the formula (D.12) and variables S and V. Then

$$R_{\min} = \Delta E - T\Delta S + P\Delta V + \frac{1}{2}\frac{\partial^2 E}{\partial S^2}\Delta S^2 + \frac{1}{2}\frac{\partial^2 E}{\partial V^2}\Delta V^2 + \frac{\partial^2 E}{\partial V \partial S}\Delta S\Delta V.$$

Since

$$\frac{\partial E}{\partial S} = T, \quad \frac{\partial E}{\partial V} = -P,$$

then

$$R_{\min} = \Delta E - T\Delta S + P\Delta V$$

$$+ \frac{1}{2}\left(\frac{\partial T}{\partial S}\right)_V \Delta S^2 - \frac{1}{2}\left(\frac{\partial P}{\partial V}\right)_S \Delta V^2 + \left(\frac{\partial T}{\partial V}\right)_S \Delta S\Delta V.$$

Therefore,

$$R_{\min} = \frac{1}{2} \left\{ \frac{T}{C_V} \Delta S^2 - \left(\frac{\partial P}{\partial V} \right)_S \Delta V^2 + 2 \left(\frac{\partial T}{\partial V} \right)_S \Delta S \Delta V \right\}. \qquad (D.16)$$

Let us introduce the following dimensionless constants:

$$\alpha = \frac{1}{C_V}, \quad \beta = \frac{V}{T} \left(\frac{\partial T}{\partial V} \right)_S, \quad \gamma = -\frac{V^2}{T} \left(\frac{\partial P}{\partial V} \right)_S \qquad (D.17)$$

and the dimensionless variables

$$\begin{cases} x = \Delta S \\ y = \dfrac{\Delta V}{V} \end{cases} . \qquad (D.18)$$

Then

$$W = c e^{-\frac{1}{2}(\alpha x^2 + \gamma y^2 + 2\beta xy)}. \qquad (D.19)$$

The constant c^* for a two-dimensional distribution (A3.19) is determined by means of the conventional normalization condition:

$$\int_{-\infty}^{\infty} W(x,y) dx dy = 1. \qquad (D.20)$$

Now we must diagonalize the quadratic form:

$$\Phi = \alpha x^2 + \gamma y^2 + 2\beta xy. \qquad (D.21)$$

We introduce the matrix

$$A = \begin{pmatrix} \alpha & \beta \\ \beta & \gamma \end{pmatrix}. \qquad (D.22)$$

As we know from linear algebra, matrix Λ may be presented in the form

$$A = B\Lambda B^{-1}, \qquad (D.23)$$

where $\Lambda = \begin{pmatrix} \lambda_{1_1} & 0 \\ 0 & \lambda_2 \end{pmatrix}$ and $\lambda_{1,2}$ are the eigenvalues of matrix A determined from the equation:

$$\begin{vmatrix} \alpha - \lambda & \beta \\ \beta & \gamma - \lambda \end{vmatrix} = 0.$$

Hence

$$\lambda_{1,2} = \frac{\alpha + \gamma}{2} \pm \sqrt{\left(\frac{\alpha - \gamma}{2} \right)^2 + \beta^2}. \qquad (D.24)$$

Matrix B is the matrix of the transformation from old variables x, y to new variables x', y' and

$$B = \begin{pmatrix} B_{11} & B_{12} \\ B_{21} & B_{22} \end{pmatrix}. \qquad (D.25)$$

For convenience of record we shall introduce the variables $x_i = (x_1, x_2)$, where $x_1 \equiv x$, $x_2 \equiv y$, and the new variable $x_i' = (x_1', x_2')$. Then the transformation of "coordinates" may be presented as

$$x' = Bx, \qquad (D.26)$$

or in detailed form as

$$\begin{cases} x_1' = B_{11}x_1 + B_{12}x_2, \\ x_2' = B_{21}x_1 + B_{22}x_2. \end{cases} \qquad (D.27)$$

In terms of the new variables the positively determined quadratic form (D.21) is written as

$$\Phi = \lambda_1 x_1'^2 + \lambda_2 x_2'^2. \qquad (D.28)$$

Now we shall calculate the coefficients B_{ik}. For this we shall return to equation (D.23), which we shall rewrite in the form:

$$AB = B\Lambda. \qquad (D.29)$$

Substituting here the matrices (D.22), (D.25) and the diagonal matrix Λ, we find the following system of equations:

$$\begin{cases} \alpha B_{11} + \beta B_{12} = \lambda_1 B_{11} 1. \\ \alpha B_{12} + \beta B_{22} = \lambda_2 B_{12} 2. \\ \beta B_{11} + \gamma B_{21} = \lambda_1 B_{21} 3. \\ \beta B_{12} + \gamma B_{22} = \lambda_2 B_{22} 4. \end{cases} \qquad (D.30)$$

The first of the above equations gives

$$B_{12} = \frac{\lambda_1 - \alpha}{\beta}. \qquad (D.31)$$

From the normalization condition $B_{11}^2 + B_{12}^2 = 1$ we immediately obtain

$$\begin{cases} B_{11} = \dfrac{\beta}{\sqrt{\beta^2 + (\lambda_1 - \alpha)^2}}, \\ B_{12} = \dfrac{\lambda_1 - \alpha}{\sqrt{\beta^2 + (\lambda_1 - \alpha)^2}}. \end{cases} \qquad (D.32)$$

The other two coefficients are found quite analogously as

$$\begin{cases} B_{22} = \dfrac{\beta}{\sqrt{\beta^2 + (\lambda_2 - \gamma)^2}}, \\ B_{21} = \dfrac{\lambda_2 - \gamma}{\sqrt{\beta^2 + (\lambda_2 - \gamma)^2}}. \end{cases} \qquad (D.33)$$

From (D.32) and (D.33) we can see that

$$\begin{cases} B_{11} = B_{22}, \\ B_{12} = -B_{21}. \end{cases} \tag{D.34}$$

It is easy to verify that matrix B is orthogonal and its determinant is equal to one ($\det B = 1$). The inverse of the (D.26) and (D.27) transformation is $x = B^{-1}x'$, where

$$B^{-1} = \begin{pmatrix} B_{22} & -B_{12} \\ -B_{21} & B_{11} \end{pmatrix}. \tag{D.35}$$

The obtained transformation allows us to express, for example, $\langle x^2 \rangle$ in the form:

$$\langle x^2 \rangle = \frac{\int\limits_{-\infty}^{\infty} \int (B_{11}^2 x_1'^2 + B_{12}^2 x_2'^2) e^{-\frac{(\lambda_1 x_1'^2 + \lambda_2 x_2'^2)}{2}} \, \mathrm{d}x_1' \mathrm{d}x_2'}{\int\limits_{-\infty}^{\infty} \int e^{-\frac{(\lambda_1 x_1'^2 + \lambda_2 x_2'^2)}{2}} \, \mathrm{d}x_1' \mathrm{d}x_2'}. \tag{D.36}$$

Here we have taken into account that

$$\left(\int\limits_{-\infty}^{\infty} x e^{-\alpha x^2} \mathrm{d}x \right)^2 = 0.$$

A simple calculation of Gauss's integrals leads to the formula for the desired fluctuation:

$$\langle x^2 \rangle = \frac{B_{11}^2}{\lambda_1} + \frac{B_{12}^2}{\lambda_2}. \tag{D.37}$$

If we substitute here explicit expressions for coefficients B_{11} and B_{12} from (D.32), and for $\lambda_{1,2}$ from (D.24), then after simple algebraic transformations we get

$$\langle x^2 \rangle = \frac{\gamma}{\alpha\gamma + \beta^2}. \tag{D.38}$$

Quite similarly we may obtain the expression for $\langle y^2 \rangle$ as

$$\langle y^2 \rangle = \frac{B_{21}^2}{\lambda_2} + \frac{B_{22}^2}{\lambda_1}, \tag{D.39}$$

or

$$\langle y^2 \rangle = -\frac{\alpha}{\alpha\gamma + \beta^2}. \tag{D.40}$$

Formulae (D.38) and (D.39) enable us to calculate any fluctuations in any independent variables, as well. In particular, for this example, substituting in (D.38) explicit expressions for α, β, γ from (D.17) we obtain

$$\langle \Delta S^2 \rangle = \frac{C_V}{1 + (C_V/T)(\partial T/\partial V)_S (\partial T/\partial P)_S}. \tag{D.41}$$

Since

$$\left(\frac{\partial T}{\partial V}\right)_S = -\frac{T}{C_P}\left(\frac{\partial S}{\partial V}\right)_T,$$

and

$$\left(\frac{\partial T}{\partial P}\right)_S = -\frac{T}{C_P}\left(\frac{\partial S}{\partial P}\right)_T,$$

and, in turn, the derivatives of the entropy are

$$\left(\frac{\partial S}{\partial V}\right)_T = \left(\frac{\partial P}{\partial T}\right)_V, \quad \left(\frac{\partial S}{\partial P}\right)_T = -\left(\frac{\partial V}{\partial T}\right)_P,$$

then, substituting all this in (D.41), we find as the result the general formula for the calculation of the entropy fluctuations, that is

$$\langle \Delta S^2 \rangle = \frac{C_V}{1 - (T/C_P)(\partial P/\partial T)_V(\partial V/\partial T)_P}. \tag{D.42}$$

In the case of an ideal gas, i.e. when the equation of state $PV = NT$ is applies, we easily get

$$\left(\frac{\partial P}{\partial T}\right)_V = \frac{P}{T}, \quad \left(\frac{\partial V}{\partial T}\right)_P = \frac{V}{T},$$

and therefore

$$\langle \Delta S^2 \rangle = \frac{C_V}{1 - (N/C_P)} = \frac{C_P C_V}{C_P - N} = C_P. \tag{D.43}$$

So, we are convinced that only in the case of an ideal gas does the formula (D.42) give for the entropy fluctuations a simple expression equal to the isobaric heat capacity. Moreover, the formula (D.42) is a general one.

For the fluctuations $\langle \Delta V^2 \rangle$, according to (D.40) we have

$$\langle \Delta V^2 \rangle = V^2 \langle y^2 \rangle = \frac{-T^2}{C_V(\partial T/\partial V)_S^2 + T(\partial P/\partial V)_S}. \tag{D.44}$$

Again, as we know,

$$\left(\frac{\partial T}{\partial V}\right)_S = -\frac{T}{C_V}\left(\frac{\partial P}{\partial T}\right)_V,$$

and

$$\left(\frac{\partial P}{\partial V}\right)_S = \left(\frac{\partial P}{\partial V}\right)_T - \frac{T}{C_V}\left(\frac{\partial P}{\partial T}\right)_V^2, \tag{D.45}$$

and therefore we readily find that

$$\langle V^2 \rangle = -\frac{T^2}{(T^2/C_V)(\partial P/\partial T)_V^2 + T(\partial P/\partial V)_T - (T^2/C_V)(\partial P/\partial T)_V^2}$$

$$= -T\left(\frac{\partial V}{\partial P}\right)_T. \tag{D.46}$$

With the last formula (D.46) we conclude our review of the theory of fluctuations. We only want to stress that the calculation of *any* fluctuations and *in any* variables must be based on the application of the formulae (D.38) and (D.39). In order to use them one should choose corresponding convenient variables and use for $R_{\min}$ the expressions (D.13)–(D.15). It is obvious that the coefficients α, β and γ should be slightly modified, depending on the choice of the variables whose fluctuations we want to calculate.

And there is one more thing. With the presence of the third independent variable in the thermodynamic potentials, whose role is always played by the variable for the number of particles, the task of diagonalizing a quadratic form will be somewhat complicated, and in this case we shall have to solve the system of nine linear equations with corresponding conditions of normalization and orthogonalization. The computational algorithm, however, will remain the same as in the above derivation.

References

1. K. Gess: *Chemistry of Cellulose and its Satellites* (Leningrad 1934)
2. J.D. Bernal, J. Mason: Nature **188**, 910 (1960)
3. G.D. Scott: Nature **194**, 956 (1962)
4. J.D. Bernal, S.V. King: Discuss. Faraday Soc. **43**, 60 (1967)
5. M. Oda: Soils and Foundations **17**, 29 (1977)
6. L.Y. Gradus: *Manual of Dispersion Analysis by the Microscopy Method.* (Moscow, 1979)
7. V.A. Ivanov, V.V. Moshev, V.A. Shishkin: *Structural Mechanics of Inhomogeneous Media* (UNTS Acad. Sci. USSR 1982, Sverdlovsk) pp. 68–73
8. V.A. Ivanov: *Structural Mechanics of Composite Materials* (Sverdlovsk, 1983)
9. O.K. Garshin, G.I. Hite: *Structural Mechanics of Composite Materials* (Sverdlovsk 1983) pp. 49–52
10. L.D. Landau, E.M. Lifshitz: *Hydrodynamics*, Vol. 6 (Science, Moscow 1988)
11. L.D. Landau, E.M. Lifshitz: *Statistical Physics*, Vol. 5 (Science, Moscow 1976)
12. J.M. Ziman: *Principles of the Theory of Solids*, 2d edn. (Cambridge 1972)
13. L.D. Landau, E.M. Lifshitz: *Statistical Physics*, Vol. 9 (Science, Moscow 1978)
14. J.S. Barash: *Van der Waals Forces* (Science, Moscow 1988) pp. 123–139
15. L.N. Vaisman: *The Structure of Paper and Methods of its Control* (Forest Production, Moscow 1973) pp. 16–25
16. *Analyzer of Fiber Dimensions FDSV:* (Ref. Inform. Cellulose, Paper and Cardboard 1977) pp. 3–15
17. B.P. Eryhov: *Non-destructive Investigation Methods of Cellulose-Paper and Wood Materials* (Forest Production, Moscow 1977) pp. 184–193
18. V.N. Nezhin, V.S. Simonov, V.A. Skachkov et al.: *Structural Mechanics of Composite Materials* (Sverdlovsk 1983), pp. 42–48
19. Destruction, Vol. 1–2: *Mathematical Fundamentals of Destruction Theory* (Moscow 1975) pp. 126–257
20. V.A. Parsegian: J. Chem. Phys. **56**, 4393–4399 (1972)
21. L.D. Landau, E.M. Lifshitz: *Theory of Elasticity*, Vol. 7 (Moscow 1987)
22. L.I. Sedov: *Mechanics of Continuous Media* (Moscow 1973)
23. G.I. Skanavi: *Physics of Dielectrics: Region of Strong Fields* (Moscow 1958)
24. A.A. Abrikosov: *Introduction to the Theory of Normal Metals* (Moscow 1972)
25. I.M. Lifshitz, M.Y. Azbel, M.I. Kaganov: *Electron Theory of Metals* (Moscow 1971)
26. S. Whitehead: *Dielectric Breakdown in Solids* (Oxford 1951)
27. P. Earhart: Phil. Mag. **1**, 147 (1901)
28. B. Cross: Brit. J. Appl. Phys. **1**, 259 (1950)
29. A.P. Aleksandrov, A.F. Ioffe: St. In. Theor. Phys. **3**, 32 (1933)

30. V.T. Renne: Electricity **5**, 16 (1950)
31. B.J. Wagner: AIEE **41**, 1034 (1922)
32. S. Karman: Arch. Elektr. **13**, 174 (1924)
33. P. Dzeryfus: Bull. SEVH. **7**, 12 (1924)
34. R.W. Rjgjwski: Arch. Elektr. **13**, 153 (1924)
35. V.A. Fok: Proceedings L. Ph. TL. **5**, 52 (1928)
36. S.O. Gladkov: Physica B **161**, 237 (1990)
37. L.D. Landau, E.M. Lifshitz: *Electrodynamics of Continuous Media* (Moscow 1982)
38. S.O. Gladkov: Phys. Lett. A **149**, 388 (1990)
39. S.O. Gladkov: Physica B **167**, 159 (1990)
40. S.O. Gladkov: Phys. Lett. A **161**, 559 (1992)
41. S.O. Gladkov Phys. Lett. A **180**, 183 (1993)
42. A.F. Valter: *Physics of Dielectrics* (Moscow 1932)
43. K. Keller: Physica **17**, 511 (1951)
44. J. Vermer: Phycica **20**, 313 (1954)
45. V.A. Chuenkov: Successes of Physical Science **54**, 185 (1954)
46. V.N. Korolev, T.J. Bazhenov: Successes of Physical Science **54**, 228 (1954)
47. V.V. Krasnopevtsev, G.I. Skanavi, E.A. Konorova: Successes of Physical Science **54**, 22 (1954)
48. S.N. Koikov, A.M. Tsikin: *Physics of Dielectrics. Publications of the All-Union Conference on Physics of Dielectrics* (Moscow 1960) pp. 230–242
49. S. Sapieha, R. Seth, L. Laporette: Sven. Papperstidning **87**, 127 (1984)
50. S. Radhakrishnan: Polymer Commun. **26**, 153 (1985)
51. F.G. Blatt: *Theory of Mobility of Electrons in Solids* (New York 1957)
52. V.L. Gurevitch: *Kinetics of Phonon Systems* (Moscow 1980)
53. V.F. Gantmaher, I.B. Levinson: *Scattering of Current Carriers in Metals and Semiconductors* (Science, Moscow 1984)
54. S.O. Gladkov: Phys. Rep. **182**, 211 (1989)
55. L.D. Landau, E.M. Lifshitz: *Quantum Mechanics*, Vol. 3 (Science, Moscow 1974)
56. H. Bateman, A. Erdelyi: *Higher Transcendental Functions*, Vol. 2. (New York, Toronto, London 1953)
57. A.I. Perepechko, V.A. Grechishkin: *Application of Ultrasound to Substance Investigation* (Moscow State University, Moscow 1971) pp. 299–301
58. V.G. Karachentsev, N.A. Kozlov: *Proc. 4th All-Union Meeting on the Chemistry and Technology of Cellulose Derivatives* (Vladimir 1972) pp. 28–34
59. V.P. Epifanov: High-Molecular Compounds **20**, 942 (1978)
60. V.P. Epifanov: High-Molecular Compounds **21**, 710 (1978)
61. V.A. Regel, A.I. Slutsker, E.E. Tomashevsky: *Kinetic Nature of the Strength of Solids* (Moscow 1974)
62. P.A. Marukov, I.N. Yusupov, P.N. Bobojanov et al.: Reports of Academy of the Sciences of the USSR **256**, 456 (1981)
63. V.G. Kachur, R.G. Zhbankov: *Proc. 16th Annual Conference NIIHTTS* (Tashkent 1981)
64. M.V. Tsympotkina, L.K. Kolmakova, I.S. Tyukova, A.A. Tager: *Thesis of Meeting on Phys. and physical-chemical Aspects of Cellulose* (Riga 1981) pp. 35–39
65. V.I. Peskovets, V.P. Nozdrin: *Proc. All-Union Conf. on Chemistry and Physics of Cellulose* (Tashkent 1982) pp. 101–102

66. B.D. Rysyuk, B.K. Yunusov, M.P. Nosov: *Proc. All-Union Conf. on Chemistry and Physics of Cellulose* (Tashkent 1982) pp. 110–111

67. G.N. Dul'nev, J.P. Zarichnyak: *Heat Conduction of Mixtures and Composite Materials* (Leningrad 1974)

68. M.I. Kulak: *All-Union Conf. on Physics and Chemistry of Cellulose* (Minsk 1990) pp. 55–57

69. S.O. Gladkov, N.Y. Kuznetsova: *All-Union Conf. on Physics and Chemistry of Cellulose* (Minsk 1990) pp. 96–99

70. S.O. Gladkov, N.Y. Kuznetsova: *Proc. Conf. on Problems of Polymer Strength and Elasticity* (Dushanbe 1990) pp. 101–103

71. S.O. Gladkov: Physica B **167**, 75 (1990)

72. S.O. Gladkov, N.Y. Kuznetsova: *Lignocellulosics, Science, Technology, Development and Use* (Ellis Horwood, 1992) pp. 429–433

73. S.O. Gladkov, N.Y. Kuznetsova: *Lignocellulosics, Science, Technology, Development and Use* (Ellis Horwood, 1992) pp. 443–447

74. S.O. Gladkov: *Proc. of the Int. Sympos. on Fiber Science and Technology* (Yokohama 1994) pp. 135

75. H.S. Carslaw, J.C. Jaeger: *Conduction of Heat in Solids*, 2nd edn. (Oxford 1957)

76. R. Kubo: *Thermodynamics of Irreversible Processes* (Moscow 1962) pp. 345–360

77. V.B. Berestecky, E.M. Lifshitz, L.P. Pitaevsky: *Quantum Electrodynamics* (Science, Moscow 1980)

78. M.A. Lavrent'ev, B.V. Shabat: *Methods and Functions of the Theory of Complex Variable* (Science, Moscow 1973)

79. A.Z. Piriatiyanskii: J. Tech. Phys. **22**, 1556 (1952)

80. V.I. Sarafanov: J. Exp. Theor. Phys. **27**, 590 (1954)

81. G.I. Skanavi, V.I. Sarafanov: J. Exp. Theor. Phys. **27**, 595 (1954)

82. O.W.L. Richardson: (Longmans-Green, 1921)

83. S.O. Gladkov: Phys. Rep. **132**, 277 (1986)

84. L.D. Landau, E.M. Lifshitz: *Physical Kinetics*, Vol. 10 (Science, Moscow 1979)

85. A.A. Abrikosov, I.E. Dziyaloshinskii, L.P. Gor'kov: *Methods of the Quantum Theory of Fields in Statistical Physics* (Science, Moscow 1962)

86. S.O. Gladkov: *Physics of Porous Structures* (Science, Moscow 1997)

87. I.G. Assovskii, O.I. Leipunskii: Physics of combustion and explosion **16**, 3 (1980)

88. I.S. Lyubchenko, V.V. Matveev, G.N. Marchenko: Reports of the Sciences of the USSR **321**, 431 (1981)

89. A.M. Baranovskii: Physics of Combustion and Explosion **10**, 95 (1983)

90. D.A. Frank-Kamenetckii: *Diffusion and Heat Transfer in Chemical Kinetics* (Science, Moscow 1967)

91. S.O. Gladkov: Phys. Lett. A **148**, 253 (1990)

92. S.O. Gladkov: Progress in Energy and Combustion Science **21**, 199 (1995)

93. L.M. Vaisman: *Capacitor Paper* (Moscow 1985)

94. A.I. Ahiezer, V.G. Bar'yahtar, S.V. Peletminskii: *Spin Waves* (Science, Moscow 1967)

95. R.J. Hardy: Phys. Rev. **132**, 168 (1963)

96. A.A. Maradudin: J. Amer. Chem. Soc. **86**, 3404 (1964)

97. G.S. Zavt: Publication of Institute Physics and Astronomy ESSR **29**, 95 (1964)

98. R.J. Hardy, R.I. Stevensen, W.C.S. Shieve: J. Math. Phys. **6**, 1741 (1965)
99. E.J. Woll: Phys. Rev. **137**, A95 (1965)
100. A.A. Maradudin: Solid State Phys. **18**, 273 (1966)
101. P. Gluck: Phys. Lett. A **24**, 292 (1967)
102. S.O. Gladkov, N.Y. Kuznetsova: Phys. Status Solidi **b 160**, K5 (1990)
103. S. O. Gladkov, N. Ya. Kuznetsova: *Lignocellulosics, Science, Technology, Development and Use* (Ellis Horwood, 1992), pp. 441–446
104. A.E. Sheidegger: *Physics of Flow of Liquids through Porous Media* (Science, Moscow 1960)
105. J. Cappel, G. Brenner: *Hydrodynamics at Small Reynolds Numbers* (Moscow 1976)
106. L.P. Yarin, G.S. Suhov: *Fundamentals of the Theory of Two-Phase Medium Combustion* (Leningrad 1987)
107. W. Coffey, M. Evans, P. Grigolini: *Molecular Diffusion and Spectra* (New York, Chichester, Brisbane, Toronto, Singapore 1984)
108. V.N. Nikol'skii, M.D. Rozentsveig: New Academy of the Sciences of the USSR **2**, 64 (1959)
109. E.I. Haikin, E.N. Rumanov: Physics of Combustion and Explosion **11**, 671 (1975)
110. A.D. Lebedev, G.S. Suhov, L.P. Yarin: Physics of Combustion and Explosion **13**, 10 (1977)
111. E.U. Repik, J.P. Sosedko: New Academy of the Sciences of the USSR **86**, 345 (1982)
112. V.M. Voloshchuk: *Kinetic Theory of Coagulation* (Leningrad 1984)
113. H. Frohlich: *Theory of Dielectrics*, 2nd edn. (Oxford 1958)
114. V.V. Daniel: *Dielectric Relaxation* (New York 1967)
115. C.J.F. Bottcher, P. Bordenvijk: *Theory of Dielectric Polarization: Dielectrics in Time Dependent Fields* (Elsevier, Amsterdam 1978)
116. S.O. Gladkov: Solid State Commun. **94**, 787 (1995)
117. S.O. Gladkov: Chem. Phys. Lett. **174**, 636 (1990)
118. I.E. Tamm: *Fundamentals of the Theory of Electricity* (Science, Moscow 1966)
119. A.V. Tubal'tsev, A.S. Bogatyrev: *Energy and Material Saving Technological Processes in the Cellulose-Paper Industry* (Kiev 1985) pp. 45–49
120. L.N. Tkach, L.M. Vaisman, V.I. Soldatenko: *Energy and Material Saving Technological Processes in the Cellulose-Paper Industry* (Kiev 1985) pp. 52–57
121. A.V. Tubal'tsev, A.S. Bogatyrev: *Technology Improvement of Production of Technical and Container Sorts of Paper and Board* (Kiev 1986) pp. 16–22
122. A.N. Gubkin: News of High School **1**, 56 (1979)
123. H.S. Van der Hulst: *Scattering of Light by Small Particles* (Mir, Moscow 1961)
124. D. Dairmenjan: *Scattering of Light by Spherical Polydispersive Particles* (Moscow 1971)
125. D. Matthews, R. Hocker: *Mathematical Methods of Physics* (Moscow 1972)
126. S.O. Gladkov: *Physics of Composites: Thermodynamics and Dissipative Properties* (Science, Moscow 1999)
127. S.O. Gladkov: J. Tech. Phys. **67**, 8 (1997)
128. S.O. Gladkov: Physics Solid State **39**, 1622 (1997)
129. S.O. Gladkov, V.G. Nikol'skii: J. Tech. Phys. (Letters) **23**, 80 (1997)
130. S.O. Gladkov: J. Tech. Phys. (Letters) **24**, 29 (1998)
131. S.O. Gladkov: J. Tech. Phys. **69**, 31 (1999)

132. S.O. Gladkov: J. Tech. Phys. **69**, 89 (1999)
133. S.O. Gladkov: Perspective Materials **5**, 17 (1999)
134. S.O. Gladkov: J. Tech. Phys. **70**, 40 (2000)
135. S.O. Gladkov: Perspective Materials **1**, 16 (2000)
136. S.O. Gladkov: Perspective Materials **4**, 12 (2000)
137. A.M. Prohorov (Ed.): *Physical Encyclopaedia* (Moscow 1983)
138. V. Braun: *Dielectrics* (Moscow 1961)
139. C. Bottcher: *Theory of Electric Polarization* (Amsterdam 1952)
140. G.P. Gurov: *Foundation of Kinetic Theory* (Moscow 1966)
141. V.M. Voloshuk: *The Kinetic Theory of Coagulation* (Leningrad 1984)
142. V.L. Bonch-Bruevich, S.G. Kalashnikov: *Physics of Semiconductors* (Science, Moscow 1977)
143. P. Resibois, M. De Leener: *Classical Kinetic Theory of Fluids* (New York, London, Sydney, Toronto 1977)
144. A.J. Leggett, D. Ter Haar: Phys. Rev. **53**, A779 (1965)
145. V.E. Gurevich, B.I. Shklovskii: J. Exp. Theor. Phys. **53**, 1726 (1967)
146. V.G. Bar'yahtar, V.L. Sobolev: Phys. State Solids **15**, 2651 (1973)
147. V.L. Sobolev: Phys. State Solids **16**, 1238 (1974)
148. V.S. Lutovinov, V.L. Safonov: Phys. State Solids **21**, 2772 (1979)
149. S.O. Gladkov: J. Exp. Theor. Phys. **83**, 806 (1982)
150. M.I. Kaganov, V.M. Cukernik: J. Exp. Theor. Phys. **37**, 823 (1959)
151. A.E. Scheidegger. Statistical Hydrodynamics in Porous Medium. J. Appl. Phys. 1954, **25**, 994–1001
152. S. Kim, W.B. Russel. Modelling of Porous Media by Renormalization of the Stokes Equations. J. Fluid Mech. 1985, **154**, 269–286
153. A.D. Myshkis, V.G. Babskii, N.D.Kopachevskii, L.A. Slobozhanin, A.D. Tyuptsov. *Low-Gravity Fluid Mech.: Mathematical Theory of Capillary Phenomena.* Springer-Verlag 1987
154. J. Bear. *Dynamics of Fluids in Porous Media.* Dover 1988
155. P.M. Adler. *Porous Media: Geometry and Transport.* Butterworth-Heineman, Stoneham, MA. 1992
156. D.A. Nield, A. Bejan.*Convection in Porous Media.* Springer-Verlag 1992
157. M. Song, J. Choi, R. Viskanta. Upward Solidification of a Binary Solution Saturated Porous Medium. Int. Heat Mass Transfer, 1993, **36**, 3687–3695
158. M.J. Hall, J.P. Hiatt. Exit Flows from Highly Porous Media. Phys. Fluids, 1994, **6**, 469–479.
159. S.C. Lee, S. White, J.A. Grzesik. Effective Radiation Properties of Fibrous Composites Containing Spherical Particles. J. Thermoph. Heat Transfer. 1994, **8**, 400–405
160. M. Kaviany. *Principles of Convective Heat Transfer.* 1994, Springer-Verlag
161. O.A. Plumb. Convective Melting of Packed Beds. Int. J. Heat Mass Transfer. 1994, **37**, 829–836
162. W. Busing, H.-J. Bart. Thermal Conductivity Unsaturated Packed Beds – Comparison of Experimental Results and Estimation Methods. Chem. Eng. Proc. 1997, **36**, 119–132
163. Y. Taketsu, T. Mosuoka. Turbulent Phenomena in Flow Through Porous Media. J. Pouros Media. 1998, **1**, 243–251
164. M. Kaviany. *Principiles of Heat Transfer in Porous Media.* Springer-Verlag 1999
165. Z. Chen, R.E. Ewing, Z.-C. Shi. Numerical Treatment of Multiphase Flows in Porous Media. Proceedings, Beijing, China 1999. Springer-Verlag 2000

Springer Series in
MATERIALS SCIENCE

Editors: R. Hull   R. M. Osgood, Jr.   J. Parisi

Springer Series in
MATERIALS SCIENCE

Editors: R. Hull   R. M. Osgood, Jr.   J. Parisi

Printing (Computer to Plate): Saladruck Berlin
Binding: Stürtz AG, Würzburg